# Teaching for Depth

## Where Math Meets the Humanities

Edited by Dale Worsley

HEINEMANN
Portsmouth, NH

**Heinemann**
A division of Reed Elsevier Inc.
361 Hanover Street
Portsmouth, NH 03801–3912
www.heinemann.com

*Offices and agents throughout the world*

The editor and publisher wish to thank those who have generously given permission to reprint borrowed material:

"The Human Face of Mathematics: Challenging Misconceptions" by Susan H. Picker and John S. Berry is used by permission of the authors.

"Math and the Westward Expansion: How an Interdisciplinary Project Changed My Thinking" by Peter Dubno, Jr., is used by permission of the author.

Figures in "The Math of Art" are reprinted by permission of artist Caissa Douwes.

"Getting Smarter: A Seventh-Grade Class Researches and Reflects on Its Discussion Habits" by Matt Wayne is used by permission of the author.

Student prompts in "From Windex to Wildstrom" are used by permission of Susan Wildstrom.

"If $dy/dx = 4x^3 + x^2 - 12/\sqrt{2x^2 - 9}$, then" from *Asylum* by Amy Quan Barry. Copyright © 2001. Reprinted by permission of the University of Pittsburgh Press.

"Geometry" from *Selected Poems* by Rita Dove. Copyright © 1980 by Rita Dove. Published by Vintage, 1993. Reprinted by permission of the author.

"Absolute Zero" by Elizabeth Fox is used by permission of the author.

Excerpt from curriculum letter to parents by Brooke Jackson is used by permission of the author.

"Mallarmé in Tournon" by Rodger Kamenetz is reprinted by permission of the author.

Excerpt of "Mr. Norton's Wart Hog" from *Very Much Like Desire* by Diane Lefer. Copyright © 2000. Published by Carnegie Mellon University Press. Reprinted by permission of the author.

"Record of Class Discussion" scoring form by Jody Madell is reprinted by permission of the author.

"Grace to Be Said at the Supermarket" from *The Blue Swallows* by Howard Nemerov. Published by University of Chicago Press. Reprinted by permission of Margaret Nemerov.

"The Mark" from *Works and Days, Volume XXVI* by David Schubert, Quarterly Review of Literature Poetry Book Series. Reprinted by permission of the Quarterly Review of Literature.

Test reflection sheet by Cheryl Schafer is used by permission of the author.

"Palm Sunday" from *Palm Sunday* by Kurt Vonnegut. Copyright © 1981 by Kurt Vonnegut. Used by permission of Dell Publishing, a division of Random House, Inc.

"What Did I Learn in School? (A Recitation)" from *So Much to Do: Poems by Alan Ziegler* by Alan Ziegler. Copyright © 1981 by Release Press. Reprinted by permission of the author.

**Library of Congress Cataloging-in-Publication Data**
Teaching for depth : where math meets the humanities / edited by Dale Worsley.
    p.   cm.
    Includes bibliographical references and index.
    ISBN 0-325-00245-2 (acid-free paper)
    1. Mathematics—Study and teaching (Middle school).   2. Mathematics—Study and teaching (Secondary).   3. Literature in mathematics education.   I. Worsley, Dale, 1948–.

QA11.2 .T43   2002
510′.71′2—dc21          2002009729

*Consulting editor: Susan Ohanian*
*Editor: Victoria Merecki*
*Production service: Lisa Garboski, bookworks*
*Production coordination: Vicki Kasabian*
*Cover design: Joni Doherty*
*Typesetter: TechBooks*
*Manufacturing: Steve Bernier*

Printed in the United States of America on acid-free paper
06  05  04  03  02  VP  1  2  3  4  5

# Contents

# Acknowledgments

I thank the following individuals and organizations for helping us realize the promise of *Teaching for Depth: Where Math Meets the Humanities*: Elizabeth Fox, my wife, and Eleanor Worsley, my daughter, for their endorsement of the idea of this book at the beginning and their infinite understanding and patience as it developed; Teachers and Writers Collaborative, who put me in touch with Heinemann and supported the final manuscript with other resources; Susan Ohanian, who shepherded us through the proposal process; Victoria Merecki, our editor, along with Lois Bridges, Leigh Peake, and Roberta Lew at Heinemann; Sheila Tobias, who so generously supported our proposal; Susan Elliott, JoAnne Eresh, Barbara Freeouf, Dinah Gieske, Shelley Harwayne, Marsha Herman, Beth Janowitz, Bea Johnstone, Andrea Lowenkopf, John Maus, Michelle McCabe, Linda Metnetsky, Rachel Pickus, Jonathan Spear, Phyllis Tam, Lucy West, Claudia Widdingham, and other members of the community of New York City Board of Education Community School District Two, in Manhattan, all of whom supported the book in different ways; Sheila Breslaw, Mary Doctor, Erick Gordon, Carol Knauert, Steve Koss, Rob Menken, Stephen Murray, Cheryl Schafer, Yelena Weinstein, and the staff at The New York City Lab School for Collaborative Studies; Joe Cassidy, Augustina DiGiovanna, Christina DiZebba, Christine Dorosh, Audra Kirshbaum, and the staff of The Clinton School for Writers and Artists; Steve Seidel at Harvard's Graduate School of Education; and the Bard College Institute for Writing and Thinking.

# Preface

The historical isolation of mathematics from humanities studies in secondary education is no longer viable. It has certainly not produced enough students with adequate math skills. Only 38 percent of America's eighth-graders could figure out a 15 percent tip on the cost of a typical meal, according to a recent report from the National Assessment of Education Progress (Stigler and Hiebert 1999, 5). In 1998 the U.S. House of Representatives approved a bill exempting nearly 150,000 foreign skilled workers from normal immigration quotas to meet the needs of high-technology industries that were unable to find workers at home (182). These discouraging statistics are complemented by the social phenomenon that mathematics, despite its powerful contributions to culture, remains stigmatized. "Ignorance of mathematics has attained the status of social grace," said Morris Kline in his comprehensive study *Mathematics in Western Culture,* published in the 1950s (vii). It is still so today, and the scorn for mathematics is visible early in students' lives, as John Berry and Susan Picker demonstrate in their chapter, "The Human Face of Mathematics: Challenging Misconceptions." They describe how, when seventh-grade students were asked to draw their images of mathematicians, significant numbers produced pictures of "weirdos, wizards, and Einstein-like figures." Progressive educational movements often advocate the interdisciplinary approach to teaching that offers hope for addressing these problems, but the rigid separation of mathematics from the rest of the curriculum persists in most schools. Clearly, too many teachers and administrators have internalized the separation and come to accept it as natural. How can they come to think and feel differently?

In *Teaching for Depth: Where Math Meets the Humanities*, a team of teachers and staff developers from private, parochial, and public schools in the New York City area have explored this question. Building on a base of academic research and working in conformity with the New Standards, they experimented with ideas to combine the math and humanities curricula. Their conclusions are provocative. Math teachers must value process and the comprehension of concepts as much as they do correct answers. They must appreciate and encourage the sense of wonderment felt by professional and historical mathematicians as they developed their big ideas, and they must open themselves to the possibility of collaboration with humanities teachers. For their part, humanities teachers must learn to incorporate math components much as science classes do. They must begin to explore the history of math, observe the way mathematicians think, and incorporate more math literacy and activity into units of study. All must be aware of the complex components of students' literacies and of their growth as integrated, capable, and curious human beings. Such practices will not only better prepare students to compete in college and in the workplace but acknowledge and nurture their spirits as learners.

The fulfilling transformations in practice that resulted from the team's experiments are chronicled in four areas: humanities in the math classroom, interdisciplinary project design, math in the humanities classroom, and the math of engagement.

The section on humanities in the math classroom begins with math teacher Virginia (Ginny) Cerussi's chapter, "Laying the Foundation: Writing in the Math Classroom." Ginny ignites her students' enthusiasm for math with writing exercises that build community and deepen the students' understanding of math concepts as well as their own learning styles. Mathematics is transformed from an experience of isolated drudgery to one of community and fulfillment. In "The Probability of Poetry," by Matthew Szenher and me, metaphor is used to lance the pain students feel as they grapple with seemingly impossible math operations. Their metaphors are transformed into literature and their minds opened to new possibilities in math and in English. Tammy Vu, in her chapter, "'I Would Have Laughed. . .': A Math Classroom Transformed by Literature," uses passages from inspiring contemporary math writers David Berlinski and Simon Singh to engage her struggling calculus students. They learn concepts, express themselves more imaginatively, and dramatically improve their images of themselves and their classmates. In his chapter, "Every Class Is an English Class," staff developer Ian Hauser walks us through the constructivist theory that underlies the powerful transformations in Ginny's, Matthew's, and Tammy's classrooms and that ultimately supports this book's thesis. In "The Human Face of Mathematics: Challenging Misconceptions," researchers Susan Picker and John Berry discuss their study of middle school students' negative images of mathematicians. They demonstrate that not only images but also performance and even careers can be changed by meaningful contact with working mathematicians.

Engineering teacher David Hardy introduces the section on interdisciplinary project design with his chapter, "Life at Imaginary High." He demonstrates how students who are disaffected by pure mathematics can become engaged when math is applied in real-world projects, in this case the design of bridges. They mature as human beings as they discover the aesthetic beauty of math's outcomes and the nuances of its ethics. In "Math and the Westward Expansion: How an Interdisciplinary Project Changed My Thinking" eighth-grade math teacher Peter Dubno, Jr. describes how a collaboration with the history and English teachers on his grade team cemented his relationship with them and nurtured his practice.

The bridge across the curriculum to math in the humanities classroom is made by sixth-grade humanities teacher Kay Rothman and me in our chapter, "A Mathematical Correspondence Between Humanists." We challenge and clarify the definitions of mathematics and the humanities and discuss why, and under what circumstances, math belongs in the humanities classrooms. Tenth-grade humanities teacher Avram Kline courageously takes on the responsibility of teaching math concepts to enable students not only to see but to experience the full significance of their cultural studies in his chapter, "The Math of Art: Probing Design Principles of Classical Greece and Islam." As is true for all the contributors, his own understanding grows when he sets out to deepen that of his students. Similarly, sixth-grade humanities teacher Amy Samson enriches the experience of all in her classroom when she uses math skills in her study of the Middle Ages, "Scaling in Humanities: Expanding Confidence, Engagement and Understanding." Her startled students resist at first, as do so many, but soon delight in the sensation of two isolated disciplines coming sensibly together. In her chapter, "Encouraging Chaos," Elizabeth Fox explores the consequences of chaos theory for student writing in essays and poetry. Through composing and recombining lines in the cinquain poetry form, her students become viscerally aware of one of the most significant cross-disciplinary breakthroughs in twentieth-century thought.

In the section on the math of engagement three teachers transcend the content of their disciplines to study the causes and patterns of student behaviors in their classrooms. English teacher Matt Wayne measures classroom talk in his chapter, "Getting Smarter: A Seventh-Grade Class Researches and Reflects on Its Discussion Habits." As a by-product students see the immediate and practical relevance of proportional thinking and graphic representation. In his moving story of personal engagement with a brave, but struggling, eleventh grader, "The Mathematician's Apprentice," Matthew Szenher demonstrates the value of one-on-one mentoring and delivers larger lessons about teaching and appreciating students' idiosyncrasies. Sylvia Gross explores the mentoring process of young teachers in her chapter, "From Windex to Wildstrom: Conversations with My Teacher." The parallel threads of the entire book are drawn together in her account of how she applies her mentor's lessons in her Bronx math classroom.

In the conclusion I tell the story of the book's origins, describe its place in our contemporary educational and intellectual environments, look at the historical context of its ideas, and assess the effectiveness of its recommendations.

To augment the usefulness of *Teaching for Depth* I have generated, with help from Matthew Szenher and Susan Schwartz Wildstrom, an annotated bibliography of works that have the potential to further realize our mission. (The bibliography can be found on the Internet at *http://www.heinemann.com/shared/onlineresources/E00245/bibliography.pdf*) I have also assembled a section of models of literature and curriculum planning, with suggestions on how the models might be used. The samples include passages from classical historians and mathematicians, contemporary novelists and poets, teachers, and students.

Because the creation of this book had the character of a personal journey for me, I have linked its many parts in a personal way, calling my links "bridges" to emphasize how not only the subject areas but also the people have been connected on my journey. Similarly, the works in the annotated bibliography and the samples section are seen to be the brainchildren of individuals with distinctive voices.

In the big educational picture *Teaching for Depth: Where Math Meets the Humanities* is but a small work, but its authors have come to feel that its mission is of great importance. Imagine the different disciplines as islands. When you dive deep enough into the water that separates them, you find they are naturally connected by the earth beneath the ocean. It is only shallow teaching that maintains the illusion otherwise. We assume that our attempt to deepen teaching will be of value to math, English, and history teachers in schools using the New Standards or working with constructivist educational programs. We also hope that it will inspire freethinking teachers from any school and from any point on the academic continuum to transform the painful experience of curricular and personal isolation into the far more pleasurable one of curricular and personal communication. Students will understand more, perform better, and grow healthier minds if they do.

# Works Cited

Kline, M. 1953. *Mathematics in Western Culture*. New York: Oxford University Press.

Stigler, J. W., and J. Hiebert. 1999. *The Teaching Gap: Best Ideas from the World's Teachers for Improving Education in the Classroom*. New York: The Free Press.

# Bridge

I met Virginia (Ginny) Cerussi in one of the math and science staff development workshops I conduct for Teachers and Writers Collaborative. After publication of *The Art of Science Writing* (Worsley and Mayer 1989) I had been conducting them twice a year. In these sessions we explore writing as a tool for learning as well as a form of self-expression, all centered on content from math and science classrooms. The workshop is structured to allow for the gradual construction of one's intuition about a subject. I have always personally found it difficult to memorize and conceptualize without being first allowed to speculate, discover, discuss, visualize, experience, reflect, revisit—all phases of what I see as a natural learning process. The pain of trying to learn otherwise has been deep enough to spawn a real passion for teaching methods that allow students to play with ideas rather than labor over them.

In the workshop I invoke the spirits of poets and scientists to send the signal that teaching via the benign nourishment of the intuition is a significant choice, with thrillingly risky open-ended possibilities. One of the quotes I use is from *Chaos* (189), by Leo Kadanoff, a physicist:

> It's an experience like no other experience I can describe, the best thing that can happen to a scientist, realizing that something that's happened in his or her mind exactly corresponds to something that happens in nature. It's startling every time it occurs. One is surprised that a construct of one's own mind can actually be realized in the honest-to-goodness world out there. A great shock, and a great, great joy (Gleick 1987, 189).

I can't say with certainty that Ginny Cerussi was shocked and joyous after participating in the workshop, but I did hear from her a couple of years afterward. She called to invite me to conduct a similar workshop at the school where she was the head of her math department, the Academy of Mount St. Ursula, in the Bronx. I discovered there that Ginny had recently completed her master's thesis, "Using Journal Writing to Learn Mathematics with Understanding" (1993). I can say that I read it with joy, as the writing experiments in her classroom confirmed all that I felt was possible in the use of writing in the math as well the science classroom.

## Works Cited

Cerussi, V. 1993. "*Using Journal Writing to Learn Mathematics with Understanding.*" Master's thesis, Iona College, New Rochelle, N.Y.

Gleick, J. 1987. *Chaos: Making a New Science*. New York: Viking. Quoting Leo Kadanoff.

# 1

## Laying the Foundation

### *Writing in the Math Classroom*

### Virginia Cerussi

Research and experience have shown that mathematics is learned through a process of communication. Students need opportunities not just to listen and absorb but also to discuss what they have observed, why certain procedures work, or why they think their solution is correct. By providing opportunities for students to express themselves about their mathematical ideas, a teacher will become aware of misconceptions before they have a chance to take root. Mathematical ideas become tangible when words and symbols are found to express them. We owe it to our students to provide opportunities for them to write about their mathematical ideas. Having students write also gives the teacher the opportunity to become a learner and with this new knowledge to become more able to guide instruction in an effective way. Among the teachers of all the disciplines, however, teachers of mathematics are the most difficult to convince of the value of writing as a routine part of their teaching experience. The reason for this lack of enthusiasm has to do with the way that it is typically presented to them. They are given the impression that they must teach writing along with the mathematics, and this seems impossible in an already overwhelming curriculum.

A distinction needs to be made between introducing writing across the curriculum and implementing writing in the content area. The purpose of the former is to improve the quality of the writing, whereas the intent of writing in a content area such as mathematics is to improve the thinking behind the mathematics. This does not mean changing course content but instead incorporating writing strategies into the existing courses.

Using writing in the mathematics classroom:

- allows all students to participate simultaneously
- encourages students to be more precise than when communicating verbally
- opens a line of communication between student and teacher

- gives students another way to look at mathematical problems
- builds a sense of community and trust so that students are more willing to take risks
- allows students to begin to see mathematics in more human terms
- gives learning a new dimension by forcing students to understand and apply concepts rather than just remembering the "trick" to each operation
- expands the ability to understand difficult concepts and to communicate this understanding to others

A highly recommended form of learning by writing is to keep a journal or "learning log." The journal is a diary-like series of writing assignments, written in prose in response to the teacher's question, statement, or instructions, called a "writing prompt." Students are given a specified amount of time to read the prompt and to formulate and write their responses. A prompt can ask a question about specific content or merely ask for an expression of feelings about the class and how it is going. It is helpful for students to imagine that they are explaining the subject to someone with little or no background in the topic. This forces them to be clear and complete, since they cannot assume that their audience had any previous knowledge of mathematical processes. In this way they are helped to improve the specificity of their writing. It forces them to slow down their thinking, and writing about each small step facilitates comprehension. As they begin to diagnose their own misunderstandings, students take charge of their own thinking and assume responsibility for their own learning.

Before beginning to implement journal writing into the mathematics classroom, teachers should discuss with the students the purposes of keeping a journal. Being aware of the benefits at the outset will make them more enthusiastic. The teacher should also measure the advantages of when the writing should take place. Writing at the beginning of the class can turn a generally unproductive time of class into a constructive time for learning and free the teacher to take care of routine classroom details such as attendance and homework collection. It can also be used to detect what students think they know about a topic before it is formally introduced, which prepares them to be more receptive to a new concept. Ending a mathematics class with writing is a good way for the teacher to assess understanding and gather impressions of how class went, and it gives the students an opportunity to summarize the main ideas of the lesson. Students can also use the time to record any lingering questions they might have.

It is usually better not to correct students' math journal writing for grammar and spelling, since the objective of the writing is to get closer to how students think. Constant correction of writing mechanics forces students to use a narrower vocabulary and to take fewer risks. Although it is important to correct all mathematical errors, the writing should also not be graded for content. Doing this would tell the students how they should process the mathematical concepts and thus undercut one of the main advantages of journal writing—individualized learning and discovery.

To keep the channels of communication open, teachers need to read what their students write and to respond in writing, however briefly, to something in each journal. This way the teachers show that they are also participating in the writing process. This greatly reinforces the benefits of sharing ideas through writing. Students appreciate the teacher's comments and realize the teacher hears them and cares. The instructor often discovers a freshness of attitude and a renewed commitment to the teaching profession.

My own discovery of the value of writing in a mathematics class took place some years ago at Mount St. Ursula's Academy, a Catholic school in New York City. I had taught an honors class of twenty girls in Sequential Mathematics III, which is essentially an algebra II and trigonometry course. As a whole, the students seemed to understand concepts, arrived at correct answers to problems, and performed extremely well on the New York State Regents examination. I felt I had done a good job. The following year, I had the same group of students in a precalculus course, which was to build on the foundation laid the previous year. Each time I referred to work done in Sequential Mathematics III the students looked at me as if they had never heard any of the material. When they did occasionally admit to having some vague recollection, I could almost read their minds saying, "And you expect us to remember THAT from LAST YEAR?"

To relieve my frustration I decided to try journal writing. I introduced it during the teaching of a unit on matrices and determinants as a way for the students to make connections among the various methods of solving a system of linear equations—algebraically, graphically, and by the use of matrices and determinants. The class had used the algebraic and graphical approaches in previous course work, but the topics had been treated separately and in an isolated manner. I began with writing prompts to assess exactly how much connection between the methods had actually been made. I told the students the journals would not be graded, and that honest answers were being sought. They would lose points only if the journals were not with them in class. Each day during the unit I wrote a prompt on the blackboard and allowed five minutes for responses. Sometimes I assigned the writing at the beginning of the class and sometimes at the end.

Effective prompts ask students to compare, summarize, analyze, and explain. Varying the formulation of the questions is important, to address individuals' different learning styles. Some should lead to factual answers. Others need to be open-ended, inviting opinion. Some prompts should be "lead-ins" to the topic to be presented, which helps students develop an interest in the lesson to come. Others should ask for conclusions after the material has been presented to them. I was always surprised at the variety of answers given the same prompts.

The following are the writing prompts I gave to the precalculus students. Those assigned at the beginning of the class are marked with an asterisk (*).

I collected the journals once a week following the completion of writing prompts #4, #7, and #11. I reviewed them overnight, wrote appropriate comments on each entry of each journal, then returned them to the students at the next class meeting to be available for reference in that day's writing assignment. Each prompt is followed by an explanation of the reason I included it as well as an overview of students' responses. (A full set of student responses to two prompts is included at the end.)

Prompt #1: *What is the main difference between solving a system of linear equations using Gaussian elimination with back-substitution and the matrix version of solution of the same system?* I used this prompt to show that there are no major differences between the two methods and that the matrix version is more efficient because only the coefficients, and not the variables, are written. A major misconception that surfaced in the writing was that most students thought the variables were not being used to solve the system. I was able to clarify that the variables were still being used, but they were just not being written down.

Prompt #2: *What do you consider to be the main idea of today's lesson? Include at least one observation you have made regarding matrix multiplication.* I posed this question to assess whether students had the ability to summarize in a few sentences certain properties of matrix multiplication, most notably that it is not necessarily commutative. The responses to this prompt ran as hoped. Students summarized the properties of matrix multiplication, and practically all noted that it is not necessarily commutative.

*Prompt #3: *What do you think is meant by the identity matrix for multiplication? Are you able to find the identity matrix for a $2 \times 2$ matrix?* The purpose of my asking this before introducing the topic was to see if the students could relate their prior knowledge of the multiplicative identity of real numbers to the matrix operation. They were asked to find the identity matrix for a $2 \times 2$ matrix as a motivator for the lesson that was to follow. Students were able to relate their prior knowledge of the meaning of identity and apply it to matrices. When trying to discover the identity matrix for a $2 \times 2$, several tried a matrix consisting entirely of ones. When that did not work, a few gave up trying, but some students persisted and were able to find the identity matrix.

Prompt #4: *Explain the procedure for solving a system of linear equations using the inverse of the coefficient matrix. When is this an efficient method?* Asking this question was an attempt on my part to get students to analyze the procedure they had just used and to break it down into small steps. It had the dual purpose of forcing them to think about what they were doing while providing an opportunity to express themselves using correct mathematical vocabulary. Some students found it difficult to describe the procedure for solving a system of equations using the inverse of the coefficient matrix; however, most were able to recognize that it is an efficient method when there are several systems to solve having the same coefficient matrix, so that the inverse needs to be found only once.

*Prompt #5: *What would you do differently if you could take the test again?* I purposefully left this question broad so that students could interpret it in any way that they chose. I expected that some students would answer in a very specific manner, delineating their approaches to particular test questions, while others would answer more generally, describing their preparation for the test. Many said they would check their answers more carefully if they could take the test again. A few said they would study more; several would not change anything; and one girl said she would eat breakfast and go to bed earlier the night before. Students were asked to respond to this question before they received their graded tests back, and it was obvious they did not want to say too much in response.

Prompt #6: *Think of a real-world application that is similar to the process of multiplying two matrices and give an explanation of it.* I asked this question to show that real-world applications do exist and to spark the students' interest in discovering what they might be. Suggestions included such items as bank accounts and their corresponding interest rates, number of shares of stock and the price per share, number of payroll hours worked and the hourly wage, and numbers of items purchased in clothes shopping and their corresponding prices. Only one student thought that there were no real-world applications of matrix multiplication.

*Prompt #7: *Write a paragraph containing as many of these words as possible: matrix, determinant, identity, inverse, scalar, order, systems, properties.* I wanted the students to show their creativity in the mathematics classroom, where all too often instruction is directed toward the left side of the brain at the neglect of the creative, or right, side. Responses to the prompt varied greatly. Several wrote basic definitions; one just repeated the words with the notation that those topics had been studied; one wrote a poem; and three wrote very creative stories, a sample of which follows:

> *Matrices* are a lot like people, complicated but not impossible. They have their own *identities* and *properties* too. But unlike us, who are composed of organs and body tissues, they are composed of numbers. In fact, *matrices* are an array of numbers. They are related to *determinants*. They share (like most relatives) many of the same characteristics and *properties* and look a lot alike. *Determinants*, however, are just a number, not a beautiful rectangular array like their cousins. It's funny how organized they are—they have *orders* and many of them are squares. Some even have *inverses*—those are the most interesting types. It's amazing how frustratingly fun *matrices* can be!!

Prompt #8: *Write about some pitfall you would warn others to watch out for when they are working with matrices and determinants.* I hoped that by asking this question the students would become more aware of whatever they were viewing as pitfalls. Several students reminded others that when evaluating a determinant using expansion by minors they should make sure that the element in the row or column being used is preceded by the proper sign. Another suggested lining up like variables in a system of equations before forming the coefficient matrix. One student warned that the product is always zero when multiplying by zero. Two advised to be sure to write the elements correctly when copying the problem and forming the matrix. Another student felt that special attention should be paid to the fact that matrix multiplication is not necessarily commutative.

Prompt #9: *Which of the ways that you now know of to solve a system of linear equations do you prefer and why?* I asked this question to connect previous work on the solution of systems of equations using algebraic and graphical methods with the current work using matrices and determinants. The most popular methods for solving a system of equations were algebraic using either elimination or substitution, or by applying Cramer's rule. Using the inverse of the coefficient matrix was not a method of choice, as most viewed it as being too time consuming and inefficient. Several wrote that they would solve graphically if the system consisted of two equations in two unknowns.

*Prompt #10: *Explain whether or not the journal writing has become easier for you with repeated practice.* Prompt #11: *Do you think journal writing has aided you in understanding the concepts in this chapter and why do you feel this way?* I asked these last two writing prompts to help me evaluate the success of introducing journal writing into this precalculus class and whether it had aided students in grasping the concepts presented in the chapter. Fifteen students responded positively, three responses were negative, and one student was unsure if the journal writing had helped her. Most found that it had become easier with repeated practice. Excerpts from the journal responses to these two prompts are included to provide an unedited sampling of the more positive results.

> Yes, I think the journal writing has become easier. I don't think of it as such an odd concept anymore. In the beginning, my friends and I had fun with the name, math journal. Ooh! No, don't look in my math journal. I don't want you to know any of my deepest, darkest math secrets. Stay away. Now I see the purpose of this actually working—somewhat. But whatever, it's fun. I seem to be doing better this trimester than I have all year—unless your tests are getting easier. Or maybe the tests seem easier as a result of the journal writing. I don't know. But I like expressing myself and my feelings through writing so I guess it's yes. Yes, it has aided me.

> It may seem as a surprise, but I hated the journal writing in the beginning. It obligated me to think too much, write and explain what I'm doing. Three of the hardest jobs that I wouldn't associate with math—except for thinking of course . . . To conclude, yes, it has become easier. It's not as boring as I thought it would be. It's actually fun.

> Yes, the journal writing has become easier. Before I even have to write about a new topic, I think about it when before I would have just done it and not thought about what I was doing. Now, I think about it while learning it so when the time comes to write, it becomes easier. . . . A big way in which it has aided me is by it not being graded. Everything is graded and points go for any little error, which makes it nerve-wracking and stressful to take a test for me. In journal writing, I spend less time worrying about points and more time

thinking about the math, which helps me understand more clearly and in turn yields those points when a test comes.

All in all, this method of solving equations is efficient and sometimes fun, especially when one gets a right answer (although getting a right answer cannot be construed as understanding, but that's another story). . . . At first I thought, "What possible questions could (teacher) come up with?" Now I've realized that all the questions have helped me understand math more. Writing in this journal is almost like having a conversation with myself with the occasional voice-over by you. Writing is not an assignment. It is a getting-to-know-you game.

It seems like one would never really use matrices in real-world problems. With the advancement of technology, I wouldn't think someone would use this system to apply to their work. Unless, maybe if there was a blackout and all systems were shut and all you could use was a candle and your brain . . . Yes, the journal writing has become easier for me. In the beginning I wasn't sure if what I was writing was correct. Then I thought that if I didn't write what you were thinking I would get points off. Knowing that I can express my feelings without judgment of any sort makes me feel that I can write anything. It has also helped me with math and showed a little of my weak areas. . . . Now when someone asks me a question in math I don't answer something like, "Oh, you know the little thing on top of the number." I know what words to use. Since we've been writing it seems easier to express oneself if you have to talk or write about math.

Journal writing makes the transition from English (my 4th period) to math easier. I was very happy that I got to be a little creative in math, which is always straightforward and factual. Also, I don't know whether there's any relevance to this but I feel as though I understand more and I got 60/60 on the last test! Coincidence? . . . Not only has my understanding improved, but my grades as well. I'm also finishing my homework a lot faster. I haven't made my final career decision yet, but I'm leaning towards teaching. Should I enter the field—I definitely plan to use this method. Thanks!!

Obviously the journal writing had helped the students grasp the concepts. They were cooperative from the beginning and actually seemed to enjoy the new activity. Their writing definitely improved in clarity and detail over time. It appeared that some students' ability to communicate verbally and with precise mathematical language improved also. Perhaps the greatest of the effects of the journals was on me as an instructor, however. It profoundly changed my approach to teaching. Before, I had placed entirely too much emphasis on producing the correct answer. This often came at the expense of precise mathematical thinking. By reading my students' writing, it became obvious to me that someone with an incorrect answer because of some careless error in miscalculation could very well have a much better understanding of the problem than others who arrived at the correct answer by mimicking the procedure used to solve similar problems, by making two errors that canceled each other out, or by some other "lucky" means.

I did experience a few minor unforeseen problems in the use of the writing journals. Some students tended to crowd their writing, for instance, leaving too little space for me to respond. The solution was for them to begin a new page for each entry. Also, although writing at the beginning of class had its definite advantages, I sometimes had difficulty getting all the students to complete their writing in time to begin the lesson. The use of a timer to signal the end of the writing period solved this problem. I also sometimes collected the journals on the same day that a writing prompt was used at the beginning of class, clearly putting an end to the exercise. By collecting the journals often I was able to clear up misconceptions before they took hold. When I collected them on test days I could read them before grading the test and gain insight into the causes of errors. It took longer for me to cover the material in the textbook than it had in previous years, but the students' deeper understanding of the topics meant less time expended later on review.

In an address to the Institute for Writing and Thinking, James A. Van Allen, head of the Department of Physics and Astronomy at the University of Iowa, said, "I am never as clear about a matter as when I have just finished writing about it. The writing process itself produces that clarity. Indeed, I often write memoranda to myself solely for the purpose of clarifying my own thinking." The same experience has been demonstrated for my students and will be substantiated in the chapters to follow: writing is a uniquely powerful tool to aid in understanding.

Since my first experiment with journal writing I have become head of the math department at Mount St. Ursula's Academy. This has put me in the position to recommend the practice to younger teachers. I advise them to start slowly, with modest hope for success, as it takes time for both the teacher and students to get accustomed to such new approaches. Failure should be expected at least some of the time. Educators in the United Kingdom tell of the "ten percent solution" to classroom innovation, meaning that it is probably possible to successfully alter 10 percent of classroom instruction during a school year. This might serve as a target for teachers in the United States as well. To experiment for one class period every two weeks, five minutes each class, or one unit during the school year is not a hardship when the activity promises to foster so much inquiry and communication. Ten percent over several years has proved in our school to have a tremendous cumulative effect.

In conclusion I offer some additional suggestions for writing prompts, ones that can be adapted to any number of classrooms and topics:

- Write three questions you believe would be good test questions. Be sure to include answers.
- List some of the common mistakes you have made in class, on homework, on tests, etc.

- Explain how you study for tests.
- How can this class be improved for you?
- How can you be certain you have mastered the information you learned today?
- What do you do best in mathematics class?
- Make up a word and write a mathematical definition of it.
- Without using numbers, explain what is meant by . . .
- Comment on whether you like or dislike this topic and explain why.

# Bridge

Ginny's recommendations on the value of writing in the math classroom are encyclopedic. Many of her points will be reiterated later in other contexts. Her concepts that writing "allows all students to participate simultaneously" and that it "opens a line of communication between student and teacher" will be picked up again in chapters on engagement. That writing "encourages students to be more precise than when communicating verbally" will be expanded upon when we examine elements of balanced literacy in the classroom. That writing "allows students to see mathematics in more human terms" is virtually the theme of our book.

The writing of the students in Ginny's piece also reflects our theme, in that the pleasure to be found in math is a significant factor in reducing the alienation so many feel. "Matrices are a lot like people, complicated but not impossible. . . ." The student who wrote this in response to prompt #7 seemed to be experiencing pleasure. Pleasure is also implicit in the students' reflections on their journal writing: "It is not as boring as I thought it would be. It's actually fun." "In journal writing I spend less time worrying about the points and more time thinking about the math, which helps me understand more clearly and in turn yields those points when the test comes." "Writing is not an assignment. It is a getting-to-know-you game." "Coincidence? Not only has my understanding improved, but my grades as well."

That Ginny's students have been able to enjoy their math class thrills me, especially as the central tool of my own vocation has enabled them to achieve their success. "We owe it to our students to provide opportunities to them to write about their mathematical ideas," Ginny says. (See a further example of reflective writing in Cheryl Shafer's entry in the "Samples" section.) The next stop on the journey of this book takes us to another math classroom where the debt is being paid, this time in the form of poetry.

Memory is a faulty organ, with a tendency to put facts into a neatly meaningful order that probably never existed, but, as I recall, I was on my way to the lower school at The Dwight School on the West Side of Manhattan to conduct a workshop when I stopped to talk to a pensively friendly fellow outside the computer room. Matthew Szenher invited me to visit one of his classes later. The more we got to know each other, the more we realized we wanted to work together. I was just then revising the proposal for the book and we felt we could do something for it. Here, then, is the story of what happened, written together. In it you will see that the element of chance takes on dynamic dimensions indeed as writing becomes not only a tool for reflection and understanding but for the creation, in collaboration with anxiety, mathematics, and computers, of works of art.

# 2

## The Probability of Poetry

### Matthew Szenher *and* Dale Worsley

## From Math to Metaphor (Matthew Szenher)

A junior in my math class came after school for extra help. For about five minutes, I reiterated what I had said in class that day. She took notes and interrupted with a few questions. I wrote a problem that we solved together, applying what she had learned. Then I gave her an almost identical problem to do on her own. She bit her lip, focused intently on her paper, wrote equations in her notebook, and seemed to be on the right track. But soon she lifted her pencil and began to move the tip in circles, still looking at the page. Then she said in a trembling voice: "Mr. Szenher, this is impossible. I can't do it!" Her arms crossed over her stomach, and her whole body shook. She was holding back tears.

A large proportion of the seventeen students in her class experience a similar math anxiety. Why? Perhaps previous math teachers have emphasized the correct answer over an understanding of the logical process, where one misstep leads to failure. This pedagogical emphasis is understandable in light of the way colleges assess our students: the SATs. But I want my students to be able to think, as well as come up with correct answers, and this is hard when their defenses (furrowed brows, tight lips, and wavering attention) go up every time I require them to negotiate the seemingly Byzantine worlds of trigonometry, geometry, and probability.

Believing that the first steps in conquering a fear are to notice and then describe the emotion itself, I designed an activity for my students to write metaphors describing their frustrations with mathematics. I expected that this would be an engaging exercise because most of the students in this class found self-expression more appealing than math. I suggested that they imagine themselves at home doing a hairy math problem. After giving them a minute to get into this fearful frame of mind, I suggested they write three metaphors to describe their feelings. They put their heads down, became deeply engaged, and wrote. Here are six of their metaphors:

I feel very vulnerable like a small plant or a flower that tries to fight against something almost impossible.

Sometime shine but sometime dark. (from an ESL student)

I could feel the pressure as if I were a grain of corn about to burst in the microwave.

I felt the pressure of a bug being crushed by an elephant.

It feels like drowning in the waves of the ocean.

I am a volcano and I can't explode, everything stays within, thus immense frustration.

The students' fears and pressures are fully evident in these lines, in which math is revealed as huge and powerful: an ocean, an elephant, a volcano, whereas they themselves are comparatively small and weak: a grain of corn, a bug, or a lone swimmer. I wrote metaphors describing my feelings about math as well:

I feel like a sardine darting in the ocean.

I feel like a bear rumbling through a dark forest.

Interestingly, I too see math as a huge world, but in it I am an animal with some ability or power.

The metaphor exercise was like an exhalation. The students seemed to agree. One commented: "It helped me get out my frustration. Instead of getting angry, I could write a metaphor. I felt better, surprisingly." Another said: "Hearing that other people have the same fears as I is helpful. I'm less embarrassed to ask questions which I may think are stupid."

The images were so wonderfully expressive that I wanted to show them I felt they had really accomplished something. My colleague Nick Didkovsky had written a computer program to scramble sentences into statistically similar, although nonsensical, paragraphs. (Didkovsky's program is accessible via the Internet at *http://www.doctornerve.org/nerve/pages/internet/mrkvform.shtml*) I decided that, given my kids' predilection for literature, I would write a similar program to convert their metaphors into a kind of chaotic poetry.

I programmed a computer to select the word *the,* then to generate the next word by scanning the list of metaphors for every word that came after *the.* It would calculate the frequency of each word's juxtaposition with *the.* Thus, for example, *fish* followed *the* three times in the metaphor list, *pressure* followed *the* twice, *smaller* five times, and so forth. The generator could choose any of the words adjacent to *the* in the metaphors, but it was more likely to pick a word with a higher frequency rating. Thus, the poems were statistically similar to the metaphors: there was a good chance that short phrases in the metaphors would be repeated in the poems, but as the phrases got longer, this likelihood decreased. Suppose that the generator selected *pressure* to appear after *the*. To generate the third word in the poem, it would scan the metaphors for all words following *pressure* and select one. Again, the words that appeared after *pressure* most frequently had a higher chance of being selected. The program generated thirty-five words in this manner, after which it continued picking words until it selected one ending with a period, question mark, or exclamation point.

Here are three generated poems:

> the only problem I feel like taking a single grape in the answer.
> within, thus immense frustration. have been before without a different
> point of seeing
> the tube, at the waves of summer.

> the middle of corn about to burst in a small plant
> in a
> small plant or move just sit still. a melting ice cube in the middle of
> summer.

> the pressure
> as a cliff and saliva coming out the tube, at the
> snout. A Marriott Hotel. a single grape in somewhere where you
>         have been
> before
> without any success, frustration.

The poems were mostly nonsensical and contained mistakes of grammar and punctuation. Nevertheless, this program proved to be useful in the classroom. For one thing, my explanation of the machinations of the program to the class included a review of probability, which they had studied earlier in the year. The program also gave them the opportunity to do some relevant science writing, as I needed cogent explanations of the algorithm. I culled the best explanations and used them in the introduction to the program itself, which is available on the Internet at *http://www.dwight.edu/PoemGen/*.

The poems themselves engendered additional activities. Some of the students were frustrated by their meaninglessness. One student wrote: "The poems annoyed me because they made NO sense. It was all jumbled and confused. I like to read something with meaning." Other students found them stimulating. One said: "The poems from the poem generator are fun to read. It is interesting to see the words of all my classmates come together to form a collaborative piece. Some poems were funny, some did not make any sense at all, but some made a lot of sense." Dale Worsley, a writer-in-residence in the school, thought it would be useful to explore these attitudes. Although we were fast approaching the end of the year and students were busy preparing for their final exams, five of them volunteered for a two-session writing workshop to revise some of the pieces generated by the poetry generator. One student described the poetry generator as a great "creativity aide," and the following results confirm it.

# From Metaphor to Literature (Dale Worsley)

Matthew's work with his students is a good illustration of writing as a tool for understanding how students think and feel. The concept of probability had become more interesting and relevant to them, and their comprehension had been enhanced. The symbols of language had been used as a conduit to the students' feelings about the symbols of math—an elegant symmetry. If, as Wordsworth asserted, poetry is "emotion recollected in tranquility," then perhaps these students would have a more tranquil reaction to a frustrating math problem next time. Matthew's work also had strong literary potential. The original compilation of metaphors sounded like a collaborative list poem:

> I feel like a cherry flavored lollipop being gnawed on.
> I feel like a snowy day in the middle of summer.
> I feel like a melting ice cube in a bowl of chicken noodle soup.
> It feels like I am standing on the edge of a cliff and I just lost my balance.
>     I plummet.
> I am a single grape in a toaster slowly losing water.
> I feel like a pretzel.
> It feels like drowning on the waves of the ocean.

The poems generated by the computer were entertaining; sometimes the transformed metaphors were shocking, or imaginative, or suggestive: "Problem. I feel like a cockroach being smashed by lightning . . ." or "a snowy day in a flower that must be opened . . ." or "feels like an elephant on the edge of summer." The images, sometimes lyrical, sometimes abstract, sometimes amusingly nonsensical, could easily occur in more traditional poetic forms.

These protoliterary qualities of the work might have remained underdeveloped but for the strength of the students' reactions to the project. It seemed natural to set up the workshops and move the work fully into the field of literature.

In our workshops I began with a short period of freewriting to get focused, then introduced some of the poetic forms suggested by their work. We examined lines from Bill Knott's abstract poem "Nights of Naomi" and an excerpt from Gertrude Stein's *Tender Buttons* (Padgett 1987, 2), which has some of the qualities of abstract poetry. These pieces showed the students that their reactions to both the nonsensical and the imaginative qualities of the computer-generated poems had a context in modern literature. To prepare them to work with poems of their own, we studied the short poetic line in James Schuyler's work and the long line in Walt Whitman's (Padgett 1987, 100–102). I also showed them an example of my attempts to convert the computer-generated poetry into more readable work:

> the very vulnerable like running around
> a snowy day in a bug being smashed by an ant being
> gnawed on. you have no ability to get
> in a problem I can't
> hold my breath anymore.

Poem Generator

**The Very Vulnerable**

We are the very vulnerable.
We run around naked on snowy days.
We are bugs being gnawed on by ants.
We have the ability to get in a problem
but we can't get out.
We can't hold our breath . . .
we can't hold our breath anymore.

                                                    Dale Worsley

The students then went to computers, extracted what might be seen as rough drafts or prewriting from the Poem Generator, and wrote versions of their own.

One student changed the articles, the prepositions, and the order of the phrases to convert his computer-generated piece into one that had more meaning for him:

the microwave. many
others and
they all know the anger and I
am a circle
my balance.
I felt like a bug being gnawed
on.

                                                    Poem Generator

**My Balance**

I felt like a bug being gnawed
in the microwave like many others.
they all know the anger on me.
I am a circle,
like others,
in my balance.

                                                    Student

In the critique that followed, the workshop group was especially enthusiastic about what he had done with the last three lines. The students found the circle image to be geometrically as well as psychologically apropos, and they thought the phrasing was sophisticated. When they questioned his use of the preposition *on*, he defended it. He said the anger was like a weight, or burden *on* him, rather than something inside him.

Another student lost the computer-generated version of his piece but wrote a second draft that made more conventional sense than the first:

The anger I possess is
like a step outside myself.
I am a computer without any success,
having my buttons pushed to the limit.

> I am like a volcano about to erupt.
> Should I move or should I sit here,
> all the while taking in the pain.

The workshop participants found this one easier to accept at first look. The repetition of phrasing and imagery was immediately powerful, and they readily identified with the question at the end.

This student took a tack similar to the first in breaking prosaic convention for poetic effect:

> the race with wild
> tigers. being crushed by an open orange. of summer.
> by lightning. of seeing the waves of the middle of summer.
>
> Poem Generator

> the race of season:
> the waves of summer are wild with tigers.
> the middle of summer is being crushed
> by the lightning
> of
> the pressure of an orange.
>
> Student

Her classmates praised the poem for the way it opposed sensations of calm and violence, for the beauty of its visual imagery, and for the cadences, which were enhanced by the line breaks. But some found it even more mystifying than the first one. They thought the comparison of lightning to the pressure of an orange was too unrealistic and felt she could strengthen the poem by moving, or even deleting, the idea of the orange. The writer defended her idea, though, asserting that although it might make not perfect sense physically, it suggested things in its ambiguity that she felt she couldn't achieve in any other way.

In our second session we all wrote poems derived from the same computer-generated version:

> the answer. do. of a keyboard. being burned
> by lighting.
> a problem from a volcano and I can't hold my balance. I
> felt as I am a cat smelling an elephant.

And here are the student revisions:

> I am being burned by lightning
> as I reach for the answer.
> My body heats like a volcano, from
> the problems I cannot solve. If I
> can hold my balance . . .
> long enough . . .
> maybe I will catch my answer.

The answer is being burned by the lightning.
A problem from this volcano can't hold my balance.
I felt as if I were a cat smelling an elephant.
The keyboard can't help me to do anything.
A foot is stepping on me.

My life is an erupting volcano
I'm being burned by everyone
The answer I do not know
It flashes before me like a bolt of lightning
I can't hold my balance anymore
It is making me very volatile
much like a cat smelling an elephant
The answer is too elaborate to be described on a keyboard

I am as a problem.
I am with the answer.
I am for the keyboard.
I am not a cat.
I cannot hold my balance.

The answer, a cat being burned by lightning.
I can't hold my balance.
A volcano, a problem from a keyboard.
I felt as smelly as an elephant.

The students' written comments on one another's work ran as follows:

The most memorable line to me was, "The answer is being burned by the lightning." I liked the same aspect of "The answer, a cat being burned by lighting" except "the cat" takes away from the line.

The most memorable poem was the one beginning "My life is an erupting volcano" because it was like a story. That made it really easy to remember.

The most memorable poem to me was "The answer, a cat being burned by lightning" because it was very deep and I liked the style he used, and the similes he used, such as the meaning of the cat.

The one beginning "My life is an erupting volcano"—volatile. The last line was what I loved: "The answer is too elaborate to be described on a keyboard."

"My life is an erupting volcano" was remarkable because of his message of a life deteriorating (death).

The students enjoyed our two workshops, but I was curious to know exactly what they might say about the experience, so I asked them to tell me in writing. Here are some responses:

The first thing I felt when I was writing this poem was that I was really free to write and there was no need to think too much. Poetry is not a hard thing to write, which I never thought before.

> This was a new outlook for me. I think this is a great idea. Usually poetry comes straight from a person's life: his experiences, thoughts, feelings. This assignment, however, combined someone's experiences, thoughts, etc., with the ideas of scrambled metaphors.
>
> My experience in poetry has flourished. By doing the exercise I have been able to express emotions and inner thought.

The writing workshops had achieved their literary objectives well enough. The students had found their own instincts reflected in conventional forms. They had experienced the pleasures of writing. They had honed their analytical abilities. Given that the poems had come from that most unforgivingly logical of disciplines, mathematics, the debate about the necessity of logic in the poetry was especially interesting. It seemed that what might at first seem chaotic, meaningless, or nonsensical, such as the scrambled metaphors, actually contained hidden logic. Mathematically, there was the logic of combining the metaphors by the law of probability. Poetically, although the syntax was sometimes execrable, the metaphors themselves might be beautiful or evocative in some new and meaningful way.

To conclude our workshop I assigned one last writing exercise, a directed freewriting on the question, How is this activity useful in the practice of math and language arts? Here are some of the answers:

> This is useful to both because it makes you think, and anything that makes you think is useful in the future. Descartes once said, "I think, therefore I am," meaning that he exists because he thinks.
>
> I can't think of a single way that these activities might be useful in math. They can help writers in using metaphors, and on doing reflection pieces.
>
> I think this exercise is useless in math because math involves numbers, not words. In language arts this will help create knowledge of poetry and give students the chance to express themselves.
>
> The best thing about this poetry writing is that it can improve our minds to make us quicker thinkers, which will help to solve math problems more quickly. It will also help you to think of ideas for writing a big essay.

That some of the students did not see an immediate application to math studies did not disturb me. The project had started simply to help humanities students with their anxiety about math. It had accomplished that and had a rich additional payoff in language arts.

# Work Cited

Padgett, R., ed. 1987. *The Teachers & Writers Handbook of Poetic Forms*. New York: Teachers and Writers.

# Bridge

The central idea of *Teaching for Depth* might be summed up in this statement: *The teaching of math can make sense to students and its pleasures can be recovered if its isolation, its "otherness," is understood and solved.* Did Matthew and I capture the Holy Grail implied in this thesis? "[Writing poetry] is useful to [the practice of math and language arts] because it makes you think, and anything that makes you think is useful in the future." For this student math's "otherness," at least in some abstract, overarching way, seems to be understood. Pleasure has certainly been recovered. And Frederick says, "The best thing about this poetry writing is that it can improve our minds to make us quicker thinkers, which will help us to solve math problems more quickly." I would stake my life on the validity of the idea, simply because I believe it must be so. I intuit that it must be so. But I must beware my intuition, which may, after all, be inadequately developed.

What might the research say about this? Stanislas Dehaene, in *The Number Sense* (1997, 242–245), deeply explores the neurological features of mathematical ability and understanding, examining them through the lenses of history, of culture and, most importantly for us, of education. Tracing the development of thought on math as an intuitive structure in our brains, he investigates the "Platonists'" understanding of math. It is "a reality outside us," as G. H. Hardy put it. Platonism, Dehaene says, "is a prevalent belief system among mathematicians and I am convinced they really have the *feeling* of moving in an abstract landscape of numbers or figures that exists independently of their own attempts at exploring it." Yet, after his research, he ultimately sides with what he calls the "intuitionists" or "constructivists": "I believe that most mathematicians do not just manipulate symbols according to purely arbitrary rules. On the contrary, they try to capture in their theorems certain physical, numerical, geometrical and logical intuitions."

Intuitionists can trace their lineage through Henri Poincaré, who spoke about "this intuition of pure number, the only intuition which cannot deceive us"; through Richard Dedekind, who found number to be an "immediate emanation from the pure laws of thought"; through Descartes, and Pascal, who said, "this knowledge provided by our hearts and instinct is necessarily the basis on which our reasoning has to build its conclusions." Kant said, "The ultimate truth of mathematics lies in the possibility that its concepts can be constructed by the human mind."

"The empirical results of research," Dehaene said, "tend to confirm Poincaré's postulate that number belongs to the 'natural objects of thought,' the

20

innate categories according to which we apprehend the world." His research reveals:

- that the human baby is born with innate mechanisms for individuating objects and for extracting the numerosity of small sets
- that this "number sense" is also present in animals, and hence that it is independent of language and has a long evolutionary history
- that in children, numerical estimation, comparison, counting, simple addition and subtraction all emerge spontaneously without much implicit instruction
- that the inferior parietal region of both cerebral hemispheres hosts neuronal circuits dedicated to the mental manipulation of numerical quantities

"Intuition about numbers," Dehaene concludes, "is thus anchored deep in our brain." He warns against taking intuitionist views too far, in defiance of logical principles, but reinforces the educational point that math learning needs to be a process of nurturing the intuition.

Creating delight in the growth of mathematical intuition is where the humanities come in, insofar as language may be categorized as a humanist feature of our brains. When we synthesize our knowledge of mathematics into expressible understandings, we are able to communicate, and enjoy them. Language also provides the mechanism for the reiteration that builds greater math skills onto their intuitive beginnings in our brains. As neuropsychologist Brian Butterworth put it:

> Trying to express something can help you understand it better. We know from a number of studies that language and numbers occupy different regions of the brain. It's nevertheless the case that some of what we know about numbers is stored linguistically. For example, if we learned our multiplication tables by rote, they lodged in the language part of our brain as a kind of poem or perhaps even a nonsense verse, depending on how well we understood them.
>
> There is an important transmission component of learning about numbers through language. Some of our knowledge remains linguistic, but most of what we know about numbers goes to the parietal lobes and is stored in a numerical rather than a linguistic way. Trying to explain mathematical ideas does help us understand what we might know implicitly (2001, 19).

Dehaene's and Butterworth's research make it safe to say, then, that some of Matthew's students have at least caught a glimpse of our grail shimmering through a thicket of computational math lessons. The gleam from the object of our quest doesn't reach the eyes of another of his students, however, at least by her account. "I can't think of a single way that these activities might be useful in math," she says. Another says, "I think this exercise is useless in math because math involves numbers, not words." Did they miss a point? Did we? I will keep these questions in mind and continue down my path, beliefs unshaken. I really don't rely on ultimate conclusions in any case. All findings are provisional, as far as I'm concerned. This lack of fixity in ideas doesn't bother me like it does some. (The Christian fundamentalists where I grew up rented billboards to implore us to "IMPEACH EARL WARREN!" They might as well have been trying to "IMPEACH EINSTEIN!" or "STOP TIME!") No, I subscribe wholeheartedly to the sentiment that the journey is, in itself, the destination.

The next stop on the journey for our readers is the classroom of Tammy Vu. As with Ginny Cerussi, I originally met Tammy when she attended one of my math and science writing workshops. A year or so later I received a packet of circle acrostics in the mail. She had taken the exercise from our workshop and tried it out in her math class. (See her chapter for examples.) I wrote one and sent it back for her to read to her high school students:

> Can tops, cut off, roll jaggedly
> Into the street, turn, like wheels,
> Round and round then
> Loop along the interstates to
> Eternal life as the mineral soul of dumps.

I'm not going to win a Nobel Prize for that one, but it was fun to write.

Once this book was under way, I remembered Tammy and got back in touch with her. A friendship formed around the drafting of her chapter. Her school happened to be located in my old neighborhood, where I had lived my bohemian years in an apartment for which the term *studio* would be overreaching. (The bathtub was in the kitchen.) Nostalgia enveloped me like perfume whenever I visited her. The students in her school were looking at their last chance of success, most of them from the struggling minority families that shared the neighborhood with struggling artists and performers. The school offered hope through its portfolio assessment system, an alternative to New York's daunting Regents curriculum. Though that hope was shattered in the spring of 2001, when the state education commissioner rescinded waivers to the Regents curriculum, it was still visible on the faces of the students who lingered in Tammy's classroom after school when we worked together.

Tammy and I became deeper thinkers and more feeling human beings because of our conversations about her writing. At least I did, because they addressed a subject that was difficult to describe well: her own transformation as a practitioner. We finally solved the problem not by pinning it down, but—perhaps not coincidentally—by *circling* it. She stated and restated certain anecdotes, drawing different conclusions and adding new ones each time, shaping her thoughts like a potter shapes clay.

# Works Cited

Butterworth, B. 2001. "Wired for Mathematics: A Conversation with Brian Butterworth." *Educational Leadership* 59 (3).

Dehaene, S. 1997. *The Number Sense: How the Mind Creates Mathematics*. New York: Oxford University Press.

# 3

## "I Would Have Laughed ..."

### *A Math Classroom Transformed by Literature*

### Tammy Vu

"Students learn math best in highly structured classes with plenty of practice problems," a fellow teacher in my department counseled. As a novice teacher at an urban Catholic high school in New York struggling to maintain control over classes of more than thirty-five students I was desperate for any advice that would help me survive. By the middle of the year, having taken his words to heart, I seemed to be succeeding. At the start of each forty-minute period, I would shut the door to my classroom and scan the six columns of seven desks for empty seats. I could take attendance in a matter of seconds this way, as my students had assigned desks. The only time they ever left them was either to go to the bathroom or to write on the blackboard. After attendance, I would start the students on the prescribed lesson, which always included example practice problems and ended with the homework assignment. The only time students were supposed to speak was when I asked them a question. They were not allowed to talk to one another. For the most part, students paid attention, took notes, participated when I asked them to write their answer on the board, and did their homework. Yet, I couldn't help feeling uneasy about my teaching. Was I evolving into exactly the type of teacher I had set out not to be?

I had decided to forgo the more lucrative and socially prestigious career paths my peers had chosen because I believed in education's power to overcome social injustices. I was concerned that mathematical thinking, an essential skill, was not well taught in secondary schools. All too often the study of mathematics was viewed as stifling creative and insightful thinking because students were forced to follow seemingly arbitrary steps for solving uninteresting problems. I believed that math was fascinating in its logical processes and essential to good critical thinking skills. The creativity lay in finding the process, and there wasn't necessarily only one way to solve a problem. Mathematicians, like artists, strive to find the most elegant way to express an idea. This is what I wanted my students to experience with mathematics.

Nevertheless, I continued to follow my mentor's advice. To develop my students' math confidence, I focused on simple, easy-to-memorize steps for finding the correct answer. On the rare occasions when I explained the purpose of a process, such as why one adds exponents when multiplying like bases, my students complained that I was making math confusing. "Just give us the rules!" they pleaded. "It's hard enough remembering the steps. Now you want us to understand why we follow each step! That's too much!"

As the year progressed I became more and more entrenched in an implicit bargain with my students: as long as I gave them rules accompanied by practice exercises, I could rest assured that they would leave my class with a positive experience. I placated myself with the thought that I was simply being realistic. Some students were good at math and others weren't. There was little point in wasting time explaining the why of math when a student couldn't even remember directions to calculate the right answer. Explanations, I rationalized, were reserved for high-level math students. Struggling students must settle for memorizing rules. But I couldn't stop doubting this teaching methodology, and I became more anxious. Could I truly consider myself a teacher?

Finally, midway through my first year, I attended a writing workshop at Teachers and Writers Collaborative, where practicing writers work in the schools with teachers and conduct staff development workshops. It was there that I discovered a way to ignite a small spark amid all the monotony. I returned to my classroom and gave my students a fifteen-minute in-class assignment to write an acrostic on the world *circle*. At first, my students were skeptical: "Poems belong in *English* class. *Numbers* belong in math class." But I stood my ground and they complied, and the results were astounding. After much giggling and whispering, they were all too eager to share their writing with the class. For the first time all year, my students turned in rapt attention to listen to one another. I heard a student in the front row note, "Wow, I never realized Melissa was sitting in the back of this class." Here are two of the students' acrostics:

> Can anyone complete the circle of love?
> In an instant the circle would shatter into a million pieces of broken
>    hearts.
> Right before your eyes, your life would fall into an endless oblivion.
> Cursed by your life, never being able to complete the circle,
> Living in an endless rotating sphere, never being able to finish,
> Even true love . . . is a circle, a circle that will never end.

> Crazy
> Idiotic
> Round
> Circles
> Laugh.
> Everywhere

In their acrostics some students expressed their appreciation of the shape as a feature of their everyday lives. Others expressed their frustration with having to calculate different aspects of geometry. Although I questioned the direct value of this exercise in affecting my students' ability to calculate arc length, length of a chord or meaningfully contemplate the significance of mathematics in their everyday life, I appreciated their enjoyment and the sense of community fostered by this exchange.

By the end of my first year I recognized, albeit marginally, the value of reading and writing in the math curriculum. But given the large student classes, short periods, and more traditional culture of the school, I believed conditions would not allow for more progress, so I moved to East Side Community High School the next fall. ECHS was a small alternative high school, meaning it was a school permitted by the New York City Board of Education to experiment with alternative forms of assessment to be able to reach students who had proven unreachable by traditional means. We used portfolio assessment rather than the state's Regents exams to evaluate our students. At ECHS we had class sizes of twenty or fewer students and block scheduling. I hoped that the school would provide the setting I needed to realize my goals.

The philosophy of the school emphasized interdisciplinary connections and student inquiry. Because of the school's emphasis on critical thinking skills, my students were already somewhat accustomed to thinking deeply. The environment was conducive to contemplating the *why* of math steps, which I felt was integral to the kind of teaching I wanted to do. Moreover the students were used to collaborating and learning from one another. This would be enormously helpful in developing their math skills, I felt.

I immediately began to integrate the use of reading and writing into my curriculum. Because my students were able to approach math with open minds, I assigned an excerpt from David Berlinski's *A Tour of the Calculus* (1995, 130) to supplement our discussion about continuous functions:

> The concept of continuity is, like so many profound concepts, both simple and elusive, elementary and divinely enigmatic. A process is continuous if it has no gaps, no place where the process itself falls into abeyance. The flight of an eagle is an example. The great bird gathers its shoulders, pushes off from a rotted tree stump, lifts into the wind, its wings beating, soars upward on a thermal current, and then, its neck curved downward, folds its wings together and dives towards the stream below. Although in the course of flight the bird does different things, there is no moment when what it does simply lapses so that it *jumps* from one part of its aerial repertoire to another.

Through this reading, students began to see real-world phenomena, such as the flight of a bird, as continuous functions. For the rest of the year, they referred to this reading when discussing functions. One student became enamored with parabolic function and joked that he wanted to tattoo the function on to his arm. Because the parabola's base would be at his elbow, he would transform it into a straight line whenever he extended his arm. Another student was so impressed by the reading that he went to the library to borrow a copy of the book and read it for himself. He felt that reading about functions described as the path of a bird in flight and learning about derivatives by picturing a woman diving into a swimming pool gave him the power to visualize the abstract math concepts more concretely. The readings were not supplemental, he believed. Rather, they were crucial to understanding the math.

On the basis of these reactions I pushed my twelfth graders, using the Berlinski excerpts as a model, to write about math concepts and describe how they enriched their world view. The outcome surpassed my expectations. Students who had pigeonholed themselves as humanities students because they scored poorly on math tests (usually as a result of forgetting details such as negative signs or making other computational mistakes) enjoyed the assignment. It was a chance for them to be creative in a discipline they had grown to view as too rigidly structured. They were able to show deep insight and to express an appreciation of functions and derivatives that would never have been evident in merely solving a math problem:

> Even though I was introduced to them in the eleventh-grade, I didn't understand derivatives until . . . I wrote a creative piece. . . . In this piece not only do I have a good story but if anyone reads it, they will be able to solve derivative problems.

Students who traditionally excel also thrived on this more literary approach. One elaborated upon the discontinuous function by describing a trip to her friend's apartment:

> You are in the lobby, waiting for the elevator so that you can go visit your friend who lives on the twenty-fifth floor. Car A only goes from the first floor to the seventeenth floor and Car B only goes from the fifteenth floor to the thirtieth floor. When you are in the elevator you get off the seventeenth floor and walk over to Car B where you can transfer to continue to the twenty-fifth floor.

This student applied the concept of turning a discontinuous function into continuous function to informally describe her life:

> There is a straight A student [who] changes and stops what she is doing because . . . she feels empty inside. Then one day she finds somebody special and he fills in the gap she was missing all along and she starts doing better again.

Another student used limits to describe a friendship:

> I felt really bad for her because her father always abused her. I tried to comfort her, but nothing would make her feel better. Then she told me not to worry because I was never going to understand how she felt. I came from a different type of family and I never experienced the abuse and neglect she had. Weird as

it may sound, she actually brought out a limit problem her teacher had given her for over the summer. She used it to show me that, until I went through what she went through, I would never fully understand her pain.

Many students began to take themselves seriously as mathematicians as a result of the writing. One projected himself as a future professional mathematician:

> It is a sunny day in the city that never sleeps. This morning I was driving up First Avenue and Houston Street. The surrounding environment has not changed since I was last around this area. The buildings are still corroding and the streets are still unclean. . . . From across the street I see my twelfth-grade math teacher Tammy, who called me two weeks ago [asking me to give] her students a lecture on derivatives since I have become a math professor.

These essays demonstrate kinds of comprehension and interest that are not evaluated on skills-oriented math tests. All students are capable of having a conceptual understanding of mathematics and of expressing their understanding verbally. When they understand the application of a math concept, they possess the motivating factor for further learning and appreciation.

I had initially planned to use readings as models only for the purpose of viewing the world mathematically; but, through readings about mathematicians, my students actually began to see the mathematicians themselves as human beings and to realize that they also faced uncertainty in their thinking. We read about mathematician Paul Erdos in the book *The Man Who Loved Only Numbers*, by Paul Hoffman (1998). The story was particularly inspiring in its portrayal of a man who patiently worked through math problems in his quest for universal truth. Erdos' pursuit was admirable and offbeat in light of the flashy rap singers and athletes my students typically admired. This reading also inadvertently raised a host of challenging philosophical issues about reality and knowledge. One that especially captured their attention was embodied in the question, Is mathematics something that is discovered or created? One student who had always been math phobic, argued passionately for discovery. "Math, the Truth, is out there to be found," he posited, "and mathematicians are searching for this Truth." In this idea of his own, that mathematicians strive against the darkness, he began to respect his own struggles with math and to see them also as a quest for truth.

When we were studying trigonometry, despite the generally open-minded attitude toward learning in the school, I was getting asked the age-old question of frustrated students, When are we ever going to use this? My answer was to read from Simon Singh's *Fermat's Enigma* (1997). I thought it would be captivating to relate the story of Pythagoras and his cult—how they used numbers in their quest for truth. As I went on, explaining the cult's beliefs and strange practices, my exasperated students cut me off and asked what the title of the book referred to. Surprised, but pleased, I explained that Fermat's "enigma" was a theorem that had stumped mathematicians for over 300 years, until British mathematician Andrew Wiles proved it in 1993. In 1637 French mathematician Pierre de Fermat stated the theorem in the margin of his copy of Diaphantus' *Arithmetica* and added, "I have a truly marvelous demonstration of this proposition which this margin is too narrow to contain." I told the students that his theorem (known as Fermat's last theorem) was simply an extension of the equation they had already memorized: $a^2 + b^2 = c^2$. Instead of squaring the variables, Fermat posited that any variation of the exponent greater than two is never true. For example, no combination of whole numbers make the following equation true: $a^3 + b^3 = c^3$.

After I explained the theorem, students copied down the equation on their own and became animated. Some pulled out their calculators to find a number that would prove the theorem wrong. One told me he would submit the proof the next day. "How much did Andrew Wiles get paid for solving the puzzle?" one asked. "I can't believe a mathematician would spend years trying to solve a simple equation." "When are you going to teach us the proof?" It hadn't been the racier aspects of the book that had captivated the students; it had been the real stuff, the math itself, but the story had served to ignite their interest.

In order to encourage my students to recognize their own capacities as mathematicians, I had developed their humanistic appreciation of the subject by employing reading and writing. These readings reinforced the message I strove to send my students, that the study of mathematics is more than getting the "right" answer—it is the process, the mental struggle, that is most significant and edifying. Math does not have to come easily to be enjoyable. Having recognized that understanding evolves from confusion, students were willing to struggle with problems. I was pleased that my students worked in groups, comfortably discussing and debating math problems with one another. They had grown to understand the process of working through math problems and realized that renowned mathematicians feel frustrations similar to their own.

My original goal had been to teach students that math is a creative and collaborative act. The sign that I had finally begun to reach that goal arrived toward the end of my third year. I had given a class a word problem: "Every time Erik eats a hot dog, he drinks two sodas. Write an equation that relates the number of hot dogs he consumes to the number of sodas he drinks. Graph this equation." After a couple minutes of working through the problem, a student asked the class, "Which equation is it? $y = 2x$ or $x = 2y$?" I went to the board, wrote the two equations, and asked, "Which one do you think it is?" Her classmates immediately volunteered their opinions, and a lively discussion ensued. After a few minutes the question was resolved and the student thanked her classmates. "Does this make sense to you now?" I asked. "Yes, it does," she answered.

My students had come to take themselves seriously as a group of mathematicians engaged in the creative process of finding patterns and unlocking the answers within math problems. Because of this motivating force and the sense of community they felt with their classmates, they were able to develop their mathematical skills, skills they will retain long after having me as their teacher. They learned to formulate incisive questions when they didn't understand, pose the questions to their classmates, and work hard to find the answers. My East Side students did not leave my class complacently convinced that they were good at math because they had memorized rules. Rather, as one student put it, "[This year] I learned things in math I never knew existed. If you would have told me last year I was going to solve problems that looked so hard, I would have laughed and said you were crazy. It taught me that you should never think you cannot do something because if you put your head to it, you can do anything."

# Works Cited

Berlinski, D. 1995. *A Tour of the Calculus*. New York: Vintage.

Hoffman, P. 1998. *The Man Who Loved Only Numbers*. New York: Hyperion.

Singh, S. 1997. *Fermat's Enigma: The Epic Quest to Solve the World's Greatest Mathematical Problem*. New York: Doubleday.

# Bridge

"Mathematicians, like artists, strive for the most elegant way to express an idea."

"The readings were crucial to understanding the math."

"[Writing enabled them] to show deep insight and to express an appreciation of functions and derivatives that would never have been evident in merely solving a math problem."

"All students are capable of having a conceptual understanding of mathematics and of expressing their understanding verbally."

The Holy Grail is nestled in my hands. Its legendary powers of regenerating life are perfectly apparent. Math has been humanized, by writing, by literature, by a humane teacher willing to ignore advice and shape the clay of her teaching in the pattern of her ideals.

So what do you do with your itinerary when you've found the object of your quest?

I stop in the shade to think about it. Whenever in doubt a writer is well advised to return to the theme of the work underway. Is there anything in it to be clarified, elaborated, more deeply plumbed? *The teaching of math can make sense to students and its pleasures can be recovered if its isolation, its "otherness," is understood and solved.* A revision comes to mind. Why not simply say, *Math can make sense, and its pleasures can be recovered if its isolation, its "otherness," is understood and solved.*

Our next stop, the mind of staff developer Ian Hauser, explores this idea of making sense, and the otherness of math, in both personal and theoretical detail.

# 4

## Every Class Is an English Class

### Ian Hauser

Wide sidewalks run along either side of Ocean Parkway, which transects my neighborhood on its path from Coney Island to Manhattan. Multistoried apartments overlook these tree-lined strips of land. In good weather the elderly Russian immigrants who live in the surrounding apartments often descend to the sidewalk's benches or play board games on the tables. They talk, they laugh, but mostly they just sit, enjoying the fresh air and camaraderie. I often wonder about their life stories. Many have lived through some of the most significant periods of recent world history. They know what it was like to survive under a Stalinist regime. They know what the Cold War felt like on the other side of the Atlantic. They know what it is like to travel across that ocean in search of a new life in the United States, how it feels to live in a new country where they do not speak the language, and where they are the "old enemy." They know so much that I will never know.

Recently, as I was stopped by the endless traffic that crawls through the nearby neighborhood of Flatbush, I noticed an African American father of about twenty cradling his infant son in his arms as they looked out of their apartment window at the passing world. They were framed by layers of graffiti on the outside of the window, surrounded by discarded beds and bedding, twisted bicycle frames and plastic bags oozing rubbish onto the already filthy street. I was struck by the tender touch of this young man as he stroked his son's arm and pointed out something beyond my view. He was smiling and whispering something obviously amusing into his son's ear, and the infant looked into his father's eyes with love and wonder.

My life as an Australian living in a house several blocks from here, but culturally a million miles away, is very different from my neighbors'. I do not know what they know. They do not know what I know, yet we look out at the same world every day.

Just how we come to *know*, to gain knowledge, is a subject that has puzzled successive generations. With the advent of sophisticated scientific tools that enable us to look inside the workings of the brain as it processes information and *thinks*, we are gaining a clearer picture of the intricacies of the human intellect. We can identify which parts of the brain are used in different learning phases, how electronic impulses travel across the nervous structure, and when different lobes are more or less active. Still the central issue remains: How does each of us come to *know* anything?

## How Do We Know What We Know?

Although we would probably agree that we do share a common knowledge about some things, what we know about them and how we come to know about them may be very different. Von Glasersfeld (1996) uses the analogy of a bowl of cherries. He says that although we may share the bowl, we don't in fact eat the same cherries at all. You and I may know that, for example, the Earth is round. In *knowing* this, I must also know what round is, what space is, what dimension is, what the Earth is, and the list goes on. For me, the pieces of knowledge, placed together in a particular way, enable me to know that indeed the Earth is round; however, each piece of knowledge is itself constructed of other pieces of knowledge, which in turn are constructed from other pieces of knowledge, and so on. I may have gained one piece of knowledge from a trip to the top of the former World Trade Center in New York, where it was possible, on a rare clear day, to actually see the curvature of the earth. You may never have had this opportunity, yet we have both come to the same shared understanding about the roundness of the earth.

While we did not eat the same cherries, we did share them. While we did not have the same learning experiences about the earth, we share a common conclusion based on different learning.

## How Can I Let You Know What I Know?
## How Can I *Teach* You?

It is not possible for me to know what the elderly Russian immigrants sitting along Ocean Parkway know, nor is it possible for me to know what that young father in Flatbush knows. I cannot live under a Stalinist regime, just as I cannot grow up as an African American in New York. The nearest I can come to knowing what these people know is to talk to them, to read about their circumstances and to relate them to things I do know in my own life. I can try to empathize with them, but I am limited by my own experiences, my own life circumstances. I cannot live their lives, I can only live my own—I can eat only my own cherries.

"Too often," Von Glasersfeld says, "teaching strategies and procedures seem to spring from the naïve assumption that what we ourselves perceive and infer from our perceptions is there, ready-made, for the students to pick up, if only they had the will to do so. This overlooks the basic point that the way we segment the flow of our experience, and the way we relate the pieces we have isolated, is and necessarily remains an essentially subjective matter. Hence, when we intend to stimulate and enhance a student's learning, we cannot afford to forget that knowledge does not exist outside a person's mind" (Von Glasersfled 1996, 5).

In writing this article it is necessary for me to connect with the experience of my audience. I must enable you to make meaning from my text. I must conform to the conventions of the English language, and I must provide you with enough examples of my thinking to enable you to tap into your own experience to bring your own meaning to what I write. If I am successful, I will be able to touch enough shared meaning to enable you to understand what I am saying here.

In searching for the word *empathy*, which I used in an earlier paragraph, I went first to a dictionary and then to a thesaurus to be certain that this was exactly the right word to use. In a way this confirms the illusion that meanings are, after all, fixed entities that do not depend on individual usage. The dictionary, as Von Glasersfeld points out, presents definitions and examples that inevitably consist of other words, which give rise to meanings only insofar as the reader interprets them. Such interpretation can be done only in terms of the chunks of perceptual and conceptual experience the individual reader has associated with the dictionary's words. "Hence," says Von Glasersfeld, "no matter how one looks at it, an analysis of meaning always leads to individual experience and the social process of accommodating the links between words and chunks of that experience until the individual deems that they are compatible with the usage and the linguistic and behavioral responses of others" (Von Glasersfled 1996, 6).

In letting students know what we know, realizing that "knowledge does not exist outside the mind of the individual," we must reconsider traditional views of education that saw teaching as "imparting knowledge," and rethink the sort of logic that imagined students as an empty vessel into which a teacher "poured" knowledge.

## Constructivism

Constructivism is a theory of learning that has developed from the work of researchers and educationalists who have sought to better understand how we come to *know*. The theory asserts that knowledge exists within each person, is established through experience, and is nurtured in the social context of human development. The basis for this theory is not new. Bartlett (1932), Bruner (1971), Smith (1975), Vygotsky (1978), Clay (1979), Cambourne (1988), and others laid the framework for the development of this theory:

- Bartlett (1932) described a theory in which students develop a "scheme" or method of understanding based on personal experience. He theorized that learners assimilate new information, which leads to subsequent changes in their schema. That is, they change their understanding based on new information.

- Bruner (1971) sought to describe the different processes that are implicated in creative problem solving. Such processes, he concluded, vary from individual to individual and from discipline to discipline.

- Smith (1975), in making the deceptively simple statement about literacy that "we learn to read and we learn through reading, by adding to what we know already (7)," underscored a theory of learning that he saw as more dynamic than Bartlett's. Smith sees that we are living in our schema. He says that within our theory of the world we can imagine and create, testing provisional solutions to problems and examining the consequences of possible behaviors. The emphasis remains on experience, of learning through doing.

- Vygotsky (1978) believed in the social process of idea making in which we learn first through person-to-person interactions and then individually through an internalization process, which leads to deep understanding.

- Clay (1979), also in reference to reading, emphasized that the story or message is the focus of the reader's attention, and although there are many other text features that need to be constantly monitored, the search for meaning is central in what good readers do. This mirrors Smith's view that involvement in the whole process advantages the learner. The "traditional wisdom" of breaking texts down to sometimes unrecognizable components, without the central focus of making meaning, most often leads to confusion.

- Cambourne (1988) says that for lifelong learning to take place, certain conditions need to be present in the classroom. He says that learners need to be *engaged* through *immersion* in appropriate tasks, given appropriate *demonstrations* of the task by competent practitioners, allowed *responsibility* to make decisions about what, when, and how they learn, given high but realistic *expectations* about their ability to complete tasks they attempt, given freedom to *approximate* mature or "ideal" performance, provided opportunities to *employ* or *use* developing skills in real and meaningful contexts, be challenged by responses and *feedback* from knowledgeable others that both supports and informs their attempts, and provided plenty of opportunities to *reflect* on and *make explicit* what they are learning.

Although none of these researchers actually used the term *constructivism*, the theory has developed from their work and from the work of others. As a theory of teaching and learning, constructivism shares a common philosophical genesis with what we might call *traditional* approaches but seeks to develop them to their next logical stage toward the evolution of more effective and efficient classroom practice.

To better illustrate the differences between constructivist classrooms and traditional practice, Brooks and Brooks (1999) developed the following chart:

| Classroom Environments | |
| --- | --- |
| **Traditional** | **Constructivist** |
| Curriculum is presented from part to whole, with emphasis on the basic skills. | Curriculum is presented from whole to part, with the emphasis on big concepts. |
| Strict adherence to fixed curriculum is highly valued. | Pursuit of student questions is highly valued. |
| Curricular activities rely heavily on textbooks and worksheets. | Curricular activities rely heavily on primary sources and data and manipulative materials. |
| Students are viewed as "blank slates" onto which information is etched by the teacher. | Students are viewed as thinkers with emerging theories of the world. |
| Teachers generally behave in a didactic manner, disseminating information to students. | Teachers generally behave in an interactive manner, mediating the environment for students. |
| Teachers seek the correct answer to validate student learning. | Teachers seek the students' point of view in order to understand students' present conceptions for use in subsequent lessons. |
| Assessment of student learning is viewed as separate from teaching and occurs almost entirely through testing. | Assessment of student learning is interwoven with teaching and occurs through teacher observations of students at work and through student exhibitions and portfolios. |
| Students primarily work alone. | Students primarily work in groups. |

# Constructivism and Literacy

Constructivist thinking has had particular implications for literacy, in which we focus on making meaning from print. Teachers are constantly searching for real, meaningful texts to use with their students. Where necessary the "sounds" of the text are taught within words, words within sentences, and sentences within meaningful contexts. A so-called balanced approach is often used, wherein reading and writing skills are demonstrated *to* the students, rehearsed *with* them, and, finally practiced *by* the students.

Literacy teachers use different techniques at each level of diminishing support by the teacher. At the upper end of teacher support is *read aloud*, in which the teacher demonstrates the process of reading. The next lower level of support is *shared reading*, which is a dialogue about the process. The next level is called *guided reading*, which allows practice (with help) of the process. The final step is *independent reading*, which is the stage of consolidation of understanding of the process. Teachers thus read *to* the students, then *with* the students, then allow the students to read *by* themselves, forming what is often called a *to-with-by* progression.

## "Every Class Is an ENGLISH Class"

So, what do constructivism and balanced literacy have to do with mathematics?

For me the answer begins in my youth, in Australia. From an elementary school of 230 students I was thrust into a secondary school of over 600 students. This is small by today's standards but seemed huge at the time. I was an average student in all subjects, but two things conspired in this year to transform me into a passionate lover of literature with equal but opposite feelings for mathematics. First, I encountered an English teacher who had me standing on my desk, along with the rest of the class, to stretch my arms like the limbs of a tree to "feel the empty space around me." He read amazing passages from books I had not heard of, sometimes with tears streaming down his face, sometimes rocking with uncontrollable laughter, but always with passion and a sense of wonder at the craftsmanship of the words on the page. I learned to look beyond the words to the metaphor, through the grammatical structure to the voice of the author, and marveled at the vicarious lives I began to live. In this teacher's class I read, wrote, reread. I presumed to criticize the works of Steinbeck, Dickens, and Eliot and even wrote a mocking version of *Julius Caesar* called *Julie of the Seesaw*.

I had the distinct feeling of floating out of his room on many occasions, only to wake up to the reality of a mathematics class that promised column after column of problems to be solved, silent struggles with tables, and unfathomable pages of "tan" and "cosine" and algorithms. It didn't take me long to learn an important lesson—that mathematics was about tricking me. The information I needed was clouded in meaningless "bumph" language whose sole purpose was to lead me away from the solution. The answer to every question regarding mathematics, from *why* to *how* was met with a stock answer by my teacher: "because mathematicians agree." I soon quit bothering to ask questions. I also learned that if you were somehow able to answer all the problems on a particular page, as a "reward" you were given yet another page of problems to answer. It didn't take me long to realize that it was much easier to get almost, but not quite, half the problems wrong. This way, all you had to do was repeat less than half a page rather than complete a new full page of problems.

Certainly, I was confronted with two very different teachers, with two very different styles, and, no doubt, with conflicting philosophies of teaching. But through many subsequent English and mathematics classes I continued to feel that in math I needed to be on guard, ready for the trick presented either by the teacher or the textbook. These deceptions were apparently designed to sift the good mathematics students from students like me. In English classes, however, I translated the "trick" into a challenge to my thinking, which I willingly talked about and thought about with my peers and teachers. This enriched my own reading and writing experience. In math I was alone. In English I was surrounded by companions on the journey to understanding.

In Australia at the time it was possible to drop mathematics after tenth grade. Having failed eighth- and ninth-grade mathematics, I did this with glee.

Throughout my high school years I had always heard a refrain from my teachers: "Every class is an English class." At the time I believed they were referring to the need for correct grammar, spelling, and form in everything I wrote. Certainly, my grades in social studies, history, geography, and economics reflected my use of the correct form of English as much as my understanding of the concepts and recall of the relevant facts. I could only presume that mathematics was exempt from this rule. After all, in math I simply needed to put the correct numbers in the correct order or substitute them for the ubiquitous $x$ or $y$.

Now a humanities staff developer for secondary schools in Manhattan, New York, I work with teachers whose primary concern is reading and writing, both in English and history classes. In conversations with colleagues who are mathematics staff developers, it has become obvious that the changing demands of the mathematics curriculum have caused math teachers to become increasingly involved in teaching literacy as well as mathematics. The paradigm shift from thinking about mathematics as a purely cognitive process to a more experiential one has brought with it a new set of priorities for teachers of mathematics. These priorities include a deeper understanding of what were previously regarded as "English language arts" techniques to encourage better reading and writing skills. Indeed, mathematics is no longer isolated. Like all the other disciplines, it, too, is now an English class.

## Balanced Mathematics?

How do we then transfer balanced literacy process to the mathematics class? I would first look at the to-with-by components and their diminishing levels of teacher support:

1.  *Demonstration of the process* (highest level of teacher support, as in "read aloud"). Vygotsky (1978) stated that in order to perform a task, learners need first to see it performed by a competent practitioner. Demonstration

is one of the greatest tools we have in teaching. Many demonstrations over an extended period will provide support for learners who will subsequently be expected to perform the task alone.

2. *Dialogue about the process* (lower level of teacher support, as in "shared reading"). Shared reading is "a think-aloud demonstration" of reading by the teacher, often using an overhead projector so that the whole class can see the text. It is an extremely powerful technique in which the learners see and hear what usually goes on "inside the head" of a competent reader. The teacher may, for instance, demonstrate what he or she does when arriving at a difficult word (reread the sentence from the beginning to gain meaning, leave the word out and read on to gain meaning, guess at the word and read on to check that the guess was correct, use knowledge of Latin or Greek roots, and so on). Students are encouraged to ask questions, to check on what they are seeing and hearing, to ask for explanation where necessary. In a mathematics class, a similar process may see the teacher working through an example while "thinking aloud" all the steps necessary to come to a conclusion. Following the presentation, the teacher may then enter into a dialogue about some of the steps taken, the process followed, the logic applied to the problem. At all times the students are free to ask for clarification, to suggest possible alternative strategies, and to challenge the teacher's solutions to problems.

3. *Practice (with help) of the process* (lower level of teacher support, as in "guided reading"). Guided reading is small-group instruction. Typically, groups are formed in which students have similar difficulties with an aspect of making meaning from text. The teacher will watch and listen while they read, to identify the aspect of the process with which the students need support, then will teach in a minilesson for this group. Groups are not static but are formed to solve specific learning difficulties and then disbanded. In a mathematics class, this group setting could facilitate discussion, perhaps with hands-on experience and real-life examples, of a problem or process in which the teacher is ready to intervene as required to provide the critical input necessary for the group to proceed. Some groups may need very little input, whereas others may require a greater amount of the teacher's specific attention.

I would stress that this kind of homogeneous grouping is used for specific teaching purposes, not as the general norm in the classroom. As someone who spent most of his time in mathematics in the "wombats" group (a wombat is a cute but notoriously dumb Australian marsupial) always watching members of the "aces" group shoot their hands up to provide solutions to problems, I can tell you that when we place low expectations on our students, they will certainly live "down" to them. Each member of a mixed-ability group will benefit from the input and perspective of all other members of the group, especially when the group members have different levels of understanding of the process. For those who are at the emergent level, demonstration of higher-level understanding will serve as an attainable goal. For those at the higher level of understanding, the need to check their thinking and explain their reasoning will serve to clarify their ideas and lead to deeper insight.

4. *Consolidation of understanding* (lowest level of teacher support, as in "independent reading"). As Smith (1994) said, children learn to read by reading, so children learn mathematical skills by using them. Cambourne pleads for "real, meaningful experiences," so the mathematics teacher will need to look for real-life, meaningful experiences in which students can become involved to consolidate what they have learned. Whereas independent reading is done alone, so consolidating understanding of a mathematical process may be best done alone, on or off the school site. Ideally, it will be interesting and achievable. Completion of seemingly endless examples to display mastery of a task can be both boring and meaningless. (It was obviously one of the things that drove me away from mathematics in my own school life.) In everything we teach, the main objective is *transfer*. Can the students transfer what they know here and now, in the classroom, to another related set of problems at another time. In other words, do they really *know*? Have they really learned anything that will last a lifetime? True consolidation of understanding is observable when students apply what they know to new and different settings. There is not much point turning the page to new problems unless and until this stage is achieved.

A classroom environment that encourages this kind of constructivist approach will need to provide space in which groups can form, perhaps an overhead projector and screen for demonstrations (though Japanese teachers usually use a chalkboard to retain the whole "story" of the lesson in view) classroom displays rich with examples of processes and solutions, and materials and resources readily available when they are needed by students.

Objectives will need to be clearly articulated so that every student is aware of both what the long- and short-term aims are, and how students are to meet them. Rubrics are a good way for students to see what level of work is required and where they currently stand in relation to the expectations of the task.

## Planning in a Constructivist Classroom

When planning instruction for a constructivist classroom, the teacher should consider different priorities. Thinking about the needs of different learners, different "entry points" for students with differing abilities, and a sequence that is well structured but offers students the opportunity to develop their own thinking are all aspects of planning for constructivism. What follows is one attempt to make this planning explicit and may be useful for teachers embarking on this new paradigm.

| Concept to be taught | Process or concept to be learned | |
|---|---|---|

| Understanding | What do I want the students to understand about the concept/process? | |
|---|---|---|

| Structure | **Demonstration of the process** | How can I best demonstrate the process? |
|---|---|---|
| | **Dialogue about the process** | How can I best encourage dialogue about the process? |
| | **Practice (with help) of the process** | Which groupings are appropriate for practice? |
| | **Consolidation of Understanding** | What activities will best encourage consolidation? |

| Vocabulary List | How can I best teach the vocabulary? | |
|---|---|---|

| Assessment | How will I know when the students have achieved the required level of understanding about the process/concept? | |
|---|---|---|

## A Word or Two About Vocabulary

Many mathematics teachers express the concern that their students have problems reading and understanding questions and examples in order to apply their mathematical skills to the solution of a problem. They feel that the students lack the vocabulary to understand what is required of them.

This problem exists in all content-area classrooms. Language arts classrooms use a number of techniques that may prove equally useful in mathematics:

- *Visual Memory:* The strategy most commonly used by effective readers, visual memory, works because the word itself, or one like it, has been seen before. The maxim "A hard word to spell is one you don't see very often" recognizes the need for teachers to display words and concepts in their classrooms. Using this strategy lends itself to the common practice of writing the word several times to see which one "looks right."

- *Drawing on Rules and Generalizations:* Berdiansky, Cronnell, and Koehla (1969) warn that the 6000 one- and two-syllable words in the spoken language vocabulary of a six- to nine-year-old use 211 sound–symbol correspondences, governed by 166 rules with 45 exceptions. Although it would seem somewhat less than efficient to try to teach all these rules and exceptions, some people find it helpful to use little "tricks" to help them remember letter order, correct letters, etc. The rule "*i* before *e*, except after *c*" is commonly used, as are tricks like "The princi*pal* is your pal." Even meaningless mnemonics such as "My Very Elderly Mother, June, Saves Us No Pie" (Mercury, Venus, Earth, Mars, Jupiter, Saturn, Uranus, Neptune, Pluto) can be sometimes useful.

- *Drawing on Linguistic Knowledge:* This strategy has been seen to be extremely effective by many teachers. It is common practice in many language arts classrooms to see lists of Latin and Greek roots, prefixes, and suffixes displayed. Apart from knowledge of the use and placement of certain types of words (like adjectives, verbs, etc.) in sentences, a basic knowledge of the roots of English words is very useful to many students. For example, the word *subtraction* comes from a combination of two Latin words: *sub,* meaning "beneath" or "below," and *trahere,* meaning "to draw." Thus, subtraction means "to draw from beneath." This type of analysis may help some students develop a depth of understanding and meaning of the process.

- *Referring to Outside Resources:* Dictionaries, thesauri, spell checkers, and the like are all outside resources. Words and concepts displayed in the classroom and referred to often, and even the teacher and other students are valuable outside resources that should be freely available in all classrooms.

- *Phonics:* Breaking words into syllables, looking for words within words, "sounding out" the word (as long as it is easily done) are all techniques that rely on knowledge of phonics in breaking open a word to understand it. Though often the first choice for teachers when introducing a new word or concept, phonics is in fact not the strategy most commonly used by effective readers.

## Do All Words Mean the Same Thing?

Although we know that in many cases a word has a particular meaning that is defined by the words around it (I *wind* the clock on the sideboard. The *wind* is very gusty in November.), some words have a particular meaning in mathematics that may not be the general usage of the word in other subject areas. Further, some words used in mathematics have a specific meaning rather than the more literal meaning often attributed to the words in general usage. These words need to be highlighted for students using all the strategies previously mentioned.

| WORD | GENERAL MEANING | MATHEMATICAL MEANING |
|---|---|---|
| **product** | the amount or quantity produced | the number or expression resulting from multiplication |
| **divide** | range between drainage areas | to subject a number or quality to the operation of finding how many times it contains another number or quality |
| **bracket** | an overhanging member or fixture that that projects from a structure and is intended as a support | one pair of marks used as signs to indicate that two or more terms are treated as one quality |

## Teaching Students to Read Mathematical Texts

Several months ago I asked a colleague, who is a mathematics staff developer for middle and high schools, to solve a problem from an eighth-grade math textbook. I wanted to see what an expert would do when confronted with such a problem, so I opened the book to the page and watched as he performed the task. The problem was spread over three pages. The first page contained a large picture and a set of statements about a class trip. The statements gave information necessary to answer several questions, which were on page three. Page two contained a map and a graph of activity relevant to the information on page one. The first thing my colleague did was scan all the information across the three pages very quickly. He then spent some time looking at the graph and the map on page two, before reading the questions on page three. Only then did he turn to page one to read the information about the problem. Having carefully read page one, he flipped to page three and as he read each question, referred back to the graph and map on page two, and wrote the answers to the questions as he went through them on page three.

What struck me as I watched my colleague go through this process was that in over two decades of teaching in schools, I had never once taught my students how to do this. Here was an expert demonstrating how he would tackle such a problem, flipping among different pieces of information, tying together written and graphic information, synthesizing meaning across different language modalities. What a wonderful shared reading this would have been for mathematics students!

Teaching students to read graphically presented information is essential in the mathematics classroom. We know from language arts that it is possible to be a reader without being a writer, but it is impossible to be an effective writer without being a reader. Applying this knowledge to mathematics, we see that it is impossible for students to generate their own graphically organized information if they do not understand graphic information. Thus, giving students lots of opportunities to graphically organize information will assist them in understanding the many graphs, maps, tables, and such that they will encounter.

Although Smith's assertion that "we learn to read by reading" (1994) makes obvious sense, I am certain that he did not mean that children should read the same type of text over and over again. Reading across different genres is something language arts teachers value greatly in assigning class reading. There is a great difference in styles between, say, narrative texts and expository or argumentative texts. Students need to read widely across many different styles in their "learning to read by reading." Similarly, in mathematics, variety is the key. I see little to be gained by students' solving page after page of similar problems. If the student knows how to perform the task, why make him or her repeat it over and over again? If the student cannot perform the task, how can repeated failure to perform help?

Open-ended, challenging, varied tasks that are meaningful and useful keep students on task. Such challenges make use of prior knowledge and new knowledge alike. Students who are having problems need help immediately, not at the end of the chapter, or after the review quiz. While other students are working on interesting and challenging open-ended tasks those who need help can be grouped (as in guided reading) and given the support they need. The emphasis for the teaching devalues from the textbook firmly into the control of the teacher, where it belongs.

## Writing to Consolidate Learning

As Virginia Cerussi, Matthew Szenher, and Dale Worsley have already noted, one of the most exciting possibilities and new realities in current mathematics teaching is the idea of having students write to reflect on their reasoning process. Writing provides students with a way to extend and deepen their understanding of concepts. It gives them the opportunity to reflect on what they are doing and to reformulate and rethink ideas. When students commit their thoughts to writing they extend and cement their learning experiences. Teachers who allow their students to choose how they will demonstrate and interpret mathematical understandings are often surprised and delighted at the variety and depth of the responses. Teachers can use writing in mathematics as part of the ongoing assessment process. Journals, portfolios, teacher observations, anecdotal records, and student self-evaluation have as much place in the mathematics classroom as they do in language arts.

## Every Class Is an *English* Class

I remember a mathematics assignment the older of my two daughters, Amy, was given in eighth grade while we still lived in Australia: "Which is a better fit: a circle within a square, or a square within a circle? Explain." The days that followed, both in class and at home, were filled with questions, explanations, and discoveries. The resultant research paper was over ten pages of writing, graphs, and illustrations. Typically, I cannot remember the answer. That was never really important. What was important was the journey of mathematical discovery. What a radical difference Amy's experience of mathematics was in our first year in the United States! As a freshman in high school she now struggled with the boredom of relentless examples and problems that often covered material she had already mastered. I remember her saying to me, "I know how to do this stuff, doesn't he [the teacher] see that?"

Thankfully, the next year we discovered a school that took a more constructivist approach—the New York City Lab School for Collaborative Studies in Manhattan. The faculty had adopted the new ARISE curriculum. Amy's math classes immediately became interesting and meaningful again. I remember my younger daughter, Felicity, explaining to a newly arrived Australian eighth-grade student that the new mathematics was "more like English and thinking, not boring like the other math." She did caution, however, that the mathematics teachers were having a real struggle because "they don't know how to teach it yet." I presume she meant that mathematics teachers are facing the same struggle language arts, history, and *all* teachers face every day—the need to change some of our classroom practices to match the new rhetoric of student-centered, constructivist learning.

Deborah Schifter, a mathematics teacher describing her experience in changing from a traditional mathematics teacher to a constructivist one says, "Creating a teaching practice guided by constructivist principles requires a qualitative transformation of virtually every aspect of mathematics teaching. The development of a new vision is only the first step (1996, 83)." In the development of that vision, the contemporary math teacher would be wise to include what constructivist teachers of literacy know, because of the demands on our students not only to solve problems but to read the language of textbooks, tests, and the texts in life beyond school that require mathematical thinking. In this way they may come to know what practicing mathematicians know about interpreting the world. They may be able to eat the cherries and be able not only to understand the taste similarly but to describe it as well.

When I pass my neighbors as they sit on their parkway benches or peer out at the world from graffiti-framed windows I know that I will never know what they know. But I can at least imagine for a moment that if we were in the same classroom—if they were my teachers—they might have a few more tools than before to describe what they know. In constructivism, then, we find the essence of humanism: that is, the ability to see something we formerly didn't see with empathetic and understanding eyes. It would be a luxury to be able to do this, to attend my neighbours' classes in being a Caribbean or a Russian immigrant. For the students in our schools the knowing of mathematics is not a luxury, it is a necessity.

The paradigm shift to constructivism is not easy. It requires a different type of thinking about how learning takes place and what the teacher and students are doing in the classroom. It requires a different way of planning for success, and different forms of assessment in verifying success. It requires a willingness to embrace change. Education has no choice but to reinvent itself along these lines. Every student in every classroom deserves nothing less.

# Works Cited

Bartlett, F. C. 1932. *Remembering: A Case Study in Experiential and Social Psychology.* London: Cambridge University Press.

Berdinasky, B., B. Cronnell, and J. A. Koehla. 1969. "Technical Report Number 15." Southwest Regional Laboratory for Educational Research and Development.

Brooks, J. G., and M. G. Brooks. 1999. *In Search of Understanding: The Case for Constructivist Classrooms.* Alexandria, VA: Association for Supervision and Curriculum Development.

Bruner, J. S. 1971. *Toward a Theory of Education.* Cambridge, MA: Harvard University Press.

Cambourne, B. 1988. *The Whole Story: Natural Learning and the Acquisition of Literacy in the Classroom.* Auckland, N.Z.: Ashton Scholastic.

Clay, M. M. 1979. *Reading: The Patterning of Complex Behavior.* Auckland, N.Z.: Heinemann.

Learning Media. 1996. *Reading for Life: The Learner as a Reader.* Wellington, N.Z.: New Zealand Ministry of Education.

Schifter, D. 1996. "A Constructivist Perspective on Teaching and Learning Mathematics." In *Constructivism: Theory, Perspectives and Practice*, ed. C. T. Fosnot. New York: Teachers College Press.

Smith, F. 1975. *Comprehension and Learning: A Conceptual Framework for Teachers.* New York: Holt, Rinehart and Winston.

———1994. *Understanding Reading.* Hilldale, NJ: Lawrence Erlbaum Assoc.

Von Glasersfeld, E. 1996. "Introduction: Aspects of Constructivism." In *Constructivism: Theory, Perpectives and Practice*, ed. C. T. Fosnot. New York: Teachers College Press.

Vygotsky, L. 1978. *Mind in Society: The Development of Higher Psychological Processes.* Cambridge: Harvard University Press.

# Bridge

Ian warrants that, "In constructivism, then, we find the essence of humanism." Constructivism is, in fact, the theory that supports our thesis. Its very premise is to create understanding and solve the otherness of knowledge, to come to common conclusions about the taste of cherries even if we eat different ones from the bowl, as Ian, citing Von Glasersfeld, puts it. Von Glaserfeld refers to the "flow of experience," reminding me again of Dewey's "continuum of experience." To me, constructivism is to education what relativity is to science: that which was separate is now continuous. Different cherries may be tasted differently by different taste buds, but their essential cherriness can be commonly understood, just as measurements may vary by location, but the essence of gravitational law remains unchanged.

Heady stuff, when we're talking only about reading, after all.

Or are we? Chuang Tzu compared words to a fish trap: "The fish trap exists because of the fish; once you've gotten the fish you can forget the trap. Words exist because of meaning; once you've gotten the meaning, you can forget the words (Keys 1985, 64)." What is reading if not a search for meaning? Coincidentally, as I write these thoughts on May 31, 2001, an article in the *New York Times* tells me that a special panel appointed by Harold Levy, Chancellor of the New York City Board of Education, is issuing a report criticizing "constructivist math" as being less than rigorous. Do they mean less than *understood*? The panelists would apparently reduce math once again to that sequence of decontextualized skills that made Ian feel alone and "tricked."

The fact is, students need to be able to read math even when it *is* decontextualized. A sixth-grade math teacher I work with, who uses the Connected Math Project's constructivist curriculum, finds that students are frequently led astray by the language when solving simple word problems. They seem to misunderstand prepositions, and their logic suffers. In her book *Twice as Less* Eleanor Wilson Orr studied the phenomenon in depth for speakers of "black English vernacular": "In English one identifies a distance by the names of the two locations that determine it: these names are held in relation to one another and to the rest of the sentence by certain prepositions. The prepositions most commonly used are *from, to* and *between*. . . . In my experience, students who are in the habit of using these prepositions according to the conventions of standard English do not confuse location and distance; and students who confuse location and distance do not use these prepositions according to those conventions (1987, 48)." The stakes are very high in this particular instance because a whole cultural group finds itself at risk of failing to cross the bridge to success in algebra.

Civil rights activist and educator Robert Moses tackled this problem head-on in The Algebra Project (2001). He arrived at a five-step teaching and learning process that is notable for its constructivist qualities in how it takes students from physical events to a symbolic representation of those events: (1) physical event, (2) picture or model of this event, (3) intuitive (idiomatic) language description of this event, (4) a description of this event in what he calls "regimented English," and (5) a symbolic representation of the event. One of the physical events that Moses employs at one site is a trip on a subway system, where students answer such questions as, At what station do we start? Where are we going? How many stops will it take to get there? In what direction do we go? The students then write about their trip, draw murals, make three-dimensional models, make graphs, collect data, and so on. Thus they come to understand the concepts previously inaccessible to them because of what one student called, after trying to read a math text book, "just words"—some of those words being *from*, *to*, and *between*!

One key element of balanced literacy is *word study*. Ian mentioned it in relation to words that have different meanings in different curricular contexts: *product*, *divide*, and *bracket*. In my travels, I have come across two articles that greatly expand on this topic. Rheta N. Rubenstein and Randy K. Schwartz point out in *Mathematics Teacher* (2000) that "much of mathematics history is reflected in the very words that we use every day. The etymologies, or origins, of mathematics words make a rich resource for deepening students' understanding and appreciation of mathematics, history and language. They open a window onto the lively history of our science and its connections with other subjects." The word *algebra,* to cite one example that seems particularly pertinent to our discussion here, "comes from the title of an Arabic work written around 825 by the Baghdad mathematician Muhammad ibn Musa al-Khowarizmi. . . . A pair of words in the title of his Arabic text, *Hisab al-jabr w'al-muqabalah*, gave us the word *algebra,* meaning 'reunion of broken parts.'" The authors conveniently chart this and other mathematical words by their roots, their links to meaning, related words, and notes.

In the *Journal of Adolescent & Adult Literacy*, Dorothy Grant Hennings (2000) demonstrates how students may, in fact, discover the anxiety-reducing information that words themselves are part of a continuum. "[C]hildren growing up in families where Graeco–Latin derived words are rarely used come to content-area studies with a disadvantage," she says. It might also interest these students that many mathematical words are derived from the human body, or from features of the landscape. The word *digit*, for instance, means not only "numbers," but also "appendages." *Leg* indicates the side of an angle. *Arm* is the name for a side of a triangle. *Hypotenuse* means "stretched against." *Geometry* is derived from *geo*, meaning "earth," and *metron*, meaning "measure."

Knowing that words are not lonesome travelers through disconnected and alienated worlds but are, in fact, members in good standing of supportive families, struggling students will doubtless feel more connected themselves, which should allow for their learning to flourish. They should never have to suffer Ian's fate ("In math I was alone . . ."). They are no different, essentially, from mathematicians throughout history, who, as Morris Kline says in *Mathematics in Western Culture*, "reach their pinnacles of pure thought not by lifting themselves by their bootstraps but by the power of social forces" (1953, 7).

It is precisely the power of these social forces that Susan H. Picker and John S. Berry address in our next chapter.

# Works Cited

Hennings, D. G. 2000. "Contextually relevant word study: Adolescent vocabulary development across the curriculum." *Journal of Adolescent & Adult Literacy* 44 (3): 270.

Keys, K. S. 1985. *Seams*. San Francisco, CA: Formant Press.

Kline, M. 1953. *Mathematics in Western Culture*. New York: Oxford University Press.

Moses, R. P., and C. E. Cobb. 2001. *Radical Equations: Math, Literacy and Civil Rights*. Boston: Beacon.

Orr, E. W. 1987. *Twice as Less*. New York: Norton.

Rubenstein, R. N., and R. K. Schwartz. 2000. "Word Histories: Melding Mathematics and Meanings." *Mathematics Teacher* 93 (8): 664–669.

# 5

## The Human Face of Mathematics

### *Challenging Misconceptions*

### Susan H. Picker *and* John S. Berry

What image do schoolchildren have of mathematicians? Does it matter? In a recent study we conducted with 476 children in the United States, the United Kingdom, Finland, Romania, and Sweden, we found that, for all practical purposes, mathematicians are invisible to middle-school children, who, when they do have images of them, often portray them in a stereotypical way as nerds and social loners (Picker and Berry 2000, 2001).

This discovery led us to ask the following questions: Would a student who sees mathematicians in this way ever seriously consider studying mathematics? Is it possible for students' attitudes to change? Would a change in attitude toward mathematicians affect a student's attitude toward mathematics?

We felt these questions were important because research has shown that images of a particular field can affect who goes into that field (Henrion 1997), and when students see mathematics as unattractive and as populated with persons very different from themselves they will be less likely to seriously consider entering the field. Moreover, in our experience, mathematics, more than any other subject they study in school, keeps students from learning how it is used in the work of its professionals. We hoped to narrow that distance by forming a panel of mathematicians to answer students' questions about their work and their lives.

We had never heard of anything similar, and we were eager to see what would ensue, but a good deal of preparation was necessary in order to guarantee that things would go as smoothly as they eventually did.

We visited every one of the seven classes in three participating schools in New York City's Community School District Two. The students brainstormed questions for the panel. They came up with 144 questions, which included the following three from every class: What do you do? Who hires mathematicians and where do they work? and How much money do you make? Continuing our research of students' images of mathematicians, we also asked students to draw a picture of "a mathematician at work." Of these drawings, confirming our suspicions, 24.7 percent contained weirdos, wizards, missing figures, and Einstein references. This drawing task would be repeated a week or so after the panel, and pupils' freewritten reactions to the panel would serve to triangulate our data.

In addition, the students were asked to suggest guidelines for their appropriate conduct on the day of the panel, which they did, and for which they agreed to take responsibility.

We chose April, Mathematics Awareness Month, to bring together the 179 seventh-grade students with eight mathematicians.

## The Panel of Mathematicians

The panel was as diverse a group as we could find. Three of the women were from countries other than the United States: Nigeria, England, and Romania. Of the two female American mathematicians, one was of African descent. Among the three male mathematicians, two were white and American; the third was from Wales. The mathematicians ranged in age from the late twenties to late forties. In meeting with classes in preparation for the panel we had heard a student say, "I'll bet they'll all be in their forties." They clearly weren't.

One of the mathematicians was a statistician; one was a researcher and ethnologist particularly interested in the mathematics to be found in games; another was a combinatorial biologist; still another mathematician was a network specialist for a major telecommunications company; one had designed defense systems; another studied epidemics; a number of them also taught and trained teachers.

Each mathematician spoke briefly to the students about his or her work and interests, after which the students had their chance to ask questions.

In the mathematicians' short presentations, students heard about mathematical ideas that could be explored with an origami cube, how mathematicians often collaborated with colleagues thousands of miles away and therefore how social mathematics could really be, what a typical day for a mathematician might be like, and they learned that many mathematicians make a very good living at it.

Students heard mathematics described as a language and a science of patterns and heard the mathematicians speak of other interests they had in their lives, including philosophy, modeling, being a guitarist in a rock group, surfing, and dancing.

In addition, as the daughter of a female mathematician, one of the panelists was able to encourage the girls in the audience to see that being a woman was not a bar to being in the field of mathematics, something that two girls in the audience asked about in particular.

With the last mathematician to speak, students were able to hear about the cutting-edge field of mathematical biology and see connections to the science they were studying in school. Students learned that mathematicians in this field used mathematical ideas to help understand disease processes, or questions of oncology, or evolution. In an illustration of one application, students looked at graphs showing the number of cases of measles in three countries—Britain, Denmark, and Iceland—and heard the kinds of questions mathematical biologists asked and looked to answer: What is going on here? Why do we see different patterns? What are the similarities and differences in the patterns we see?

This young mathematician was the closest in age to the students and he was dressed in a manner the students would describe as trendy—a dark purple shirt out over dark jeans with work boots.

## Students Question the Panel

Students now had their opportunity to ask the questions they had brainstormed together earlier. Although at first they were hesitant, eventually many students came forward. By the end of our allotted time there were at least fifteen students still waiting to speak.

In reply to a boy who asked, "Do you find that people with exceptional mathematical skills have exceptional skills in other subjects? Do mathematicians have these other skills in common?" nearly every mathematician spoke about an interest he or she had in art or music.

## The Public's Images of Mathematicians

The question of the image held by the public of mathematicians and mathematics came up early, and one of the mathematicians admitted that she sometimes found it easier not to tell people what she did for a living because so many people have had bad experiences with mathematics in school. This led the panelists to comment on the image of the nerdy mathematician. They insisted on the reality that mathematics is fun.

When one of the students asked the panel "So would you recommend us to be mathematicians?" they replied "Certainly!" and the students applauded.

## A Student Asks About Stereotypes

Much laughter was generated when a student asked, "I was wondering what you thought about the stereotype of mathematicians—like with the hat, like they're a wizard or something?" He was told, "Like all stereotypes, never believe it!" Adding to the laughter, one of the mathematicians went backstage and came back wearing his black cowboy hat. But the students were already beginning to see that with such a diverse panel, stereotyping was impossible.

At the end of general questioning and discussion, the students were instructed to write brief reactions to the panel and to state anything that surprised them.

## Students' Reactions to the Panel

The original meaning of the word *stereotype* is derived from a French process of printing from fixed plates. According to the *American Heritage College Dictionary*, it has now come to mean a "conventional, formulaic, and oversimplified conception or image." The implication here is that mathematicians, if viewed in a formulaic and oversimplified way, will be seen as undifferentiated and as having lives vastly different from the students'. In their many comments students indicated that this was, indeed, how they had seen mathematicians until the panel presented a more accurate and realistic image of themselves as individuals and as persons not so different from the students themselves. Many students chose to write about changing their beliefs, even though there was no specific prompt to do so.

## Changing Stereotypical Images of Mathematicians

The six (of twenty-one) following students' comments are typical in how they indicate careful and critical thinking that appears to have lead them to reject a number of their previously held images:

> I enjoyed this presentation because I thought it was interesting the way that all of them had different skills, opinions and reactions. It made me keep an open mind to all the differences . . . I am glad that I came.
>
> I realized that mathematicians really aren't that nerdy, they're regular people with regular lives. I see that lots of people who are mathematicians like art. I think that I might become a mathematician because I love art and I joined the math team. This might lead me to becoming one, even though I want to be a lawyer. I liked how the panel compared games to mathematics. I also liked hearing about the careers the mathematicians were interested in when they were our ages.
>
> I was surprised because they were more interesting and less stereotypical than I thought they would be. I didn't realize that there were so many fields of math. I also didn't realize how many games involved math.
>
> I was a little surprised to hear that mathematicians can make a lot of money because I thought that mathematicians didn't make a lot of money. I also found it interesting that mathematicians could do a lot of different jobs.

I think that I've learned that math can be fun, when I used to think
that it sucked. I agree with my friend Sam as well, that mathematicians have
interesting lives. Peace!

I think that some of the mathematicians were really cool. All of the
mathematicians came off as real people, which made it easier to ask questions
(and less boring).

## Changing Views of Mathematics

Along with a change in their stereotypical images, students indicated that the
invisibility of mathematicians and mathematics had diminished:

I learned that there are a lot of mathematicians but not everyone will know
that they are one.

I learned that everything has math, you may know it or not.

This student was surprised to learn that what mathematicians do is very
different from what she had thought:

I learned that mathematicians are normal people with normal lives. What
surprised me is that what they do is totally different than what I expected. It
was actually kind of interesting, and I learned that everything has math, you
may know it or not. But lots of things revolve around math.

The image of the field as narrow and monolithic had obviously changed
because of the panel. As Emmer (1990, 91) points out: "very few people realize
that there is not just one subject 'mathematics' but a whole series of different
specializations that flow into the wide river of mathematics."

Another aspect of the narrowness of students' images of mathematics is
the belief that it cannot be associated with anything they consider to be fun.
This is why they were so surprised to find that the strategies of board games are
a part of mathematics.

Students also have an image of mathematics that defines it as a school
subject, rather than an occupation (Howson and Kahane, 1990). The panel encouraged students to reexamine this belief, as the following comments indicate.

What I enjoyed most . . . is listening to the different experiences with math
from a variety of different people. I always thought mathematics only existed
in places like classrooms but now I know it's everywhere.

I learned that there is a lot more to math than what we have learned so
far from Elementary school to now. I always thought that math was just some
regular calculations but there is more to it. Things that I never thought deal with
math actually deal with math like playing games, Mancala, and engineering.

Scores of other students in our survey demonstrated that they were seeing
mathematics in a new way, but the preceding comments are typical.

## Being Inspired by the Panel

Finally, a number of students indicated in the following sampling that they were inspired by their experience of the panel:

> I really liked the mathematics panel and what they said. It in a way inspired me more to pursue my goal, which is to be a physicist. They said it's always important to have a mentor but I have not chosen one yet.
>
> I learned a lot about mathematics around the world and I think this inspired me to work harder with my math. One thing that surprised me was how much the mathematicians work with computers. It showed how much the technology has affected math.
>
> I really found the mathematicians very interesting. And was also a little shocked when the female mathematicians said that they were not discouraged because of their gender. I might think about going into the field of mathematics because of this day. Thanks!!
>
> Because of this experience I am willing to persevere and be more interested in mathematics.

## Rethinking Stereotypes

Clearly, the panel led many students to bridge the gulf they had originally felt between themselves and mathematicians. Some days later, when they had returned to their classrooms, they were asked to draw another picture of "a mathematician at work." Whereas forty (24.7 percent) of their prepanel drawings had depicted stereotypical images of weirdos, wizards, Einstein-like figures, or had been missing an image, this style of depiction was reduced to just thirteen (7.4 percent) drawings.

Just as interesting is that whereas originally 37 percent of the girls had drawn a female mathematician, 51.1 percent of the girls now depicted a female. This indicates, we believe, a greater identification of the girls with female mathematicians.

The following are three such examples of an image change in drawings created before and after the panel. (Students' names have been changed.)

In Figure 5–1 a character with wild Einsteinian hair, who has written his own version of Einstein's famous equation, is writing on a blackboard.

In his postpanel drawing, Figure 5–2, Harry appears to have drawn one of the mathematicians down to his ponytail and jeans. Harry wrote:

> My drawing is of a mathematician working in a lab at a computer figuring out an equation. (Biology.) I think that a mathematician could be hired to do just about anything related to science or math. It depends what the person studies.

Writing about the panel, Harry said:

> I learned a lot today. One of the most interesting things I learned is that math isn't just numbers, it's many things. Biology, computers, and even games like

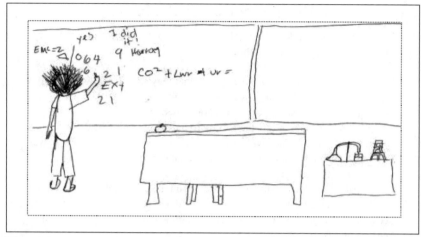

**Figure 5–1.** Harry's drawing (prepanel)

Mancala. I think that being a mathematician could be fun and exciting if you dedicated your time to it.

**Figure 5–2.** Harry's drawing (postpanel)

In Figure 5–3 we find a wild-haired figure drawn by a student named Jennifer, who wrote:

This is Albert Einstein. When I think of the word mathematician I automatically think of Albert Einstein. He is frantically writing on the board—that is what I think of him doing.

**Figure 5–3.** Jennifer's drawing (prepanel)

Writing frantically on a board is a part of the one-dimensional image many pupils have of mathematicians—that they are constantly involved in endless calculations.

In her drawing after the panel, in Figure 5–4, Jennifer not only has abandoned the Einstein and frantic writing themes but has changed the gender of the mathematician, as well.

Jennifer wrote about her second drawing:

> This is a normal everyday person. This picture also happens to be a woman. I don't mean to imply that only women can be mathematicians, however I mean to imply that women are mathematicians just as much as men are. I also noticed at the panel that many people had lots of papers with evidence on it. Also they had 3-D shapes with them.

The papers and three-dimensional shapes are represented in Figure 5–4 on the desk next to the attractive-looking woman. Jennifer wrote after the panel:

> I was surprised that they were SO INTERESTED in their field, and things like music. I don't mean that it is bad to be interested, but they were very devoted. They find ways to work 24 hours. This got me to understand what a mathematician really does and how they go about doing problems. It interested me that they took a lot of time to find a problem.

In the last set of drawings, by a student named Marnie, we again see a change in the gender of the mathematician in the second drawing. The small details in the drawing are also, we believe, quite significant.

**Figure 5–4.** Jennifer's drawing (postpanel)

In the early drawing, Figure 5–5, we see a bald mathematician with glasses doing calculations. The walls are bare, with no frame of reference or context for the figure. We believe this is because the pupils had very little sense of context for mathematicians before the panel.

The details in Marnie's postpanel depiction in Figure 5–6 are different. Although the basic stance of the two figures is the same, the female mathematician is attractive-looking and without glasses. (Of the five women on the panel, two did not wear glasses; of the three men, one did not.) The figure is situated in a more realistic setting, with mathematical objects on her desk and a computer nearby. There are now three windows in the picture, which seems to place her in the world rather than apart from it. Simple division problems are also present in this drawing, although they appear on the computer screen. And the items on the desk, which include two geometric shapes that appear to come from Dana Frank's talk, indicate that a mathematician is involved with more than just calculations. There are a pen and paper as well, but instead of calculations, the paper is inscribed, "If there were . . . ."

Annie wrote about her second drawing: "It is of a mathematician in a nice room with lots of different kinds of math around her."

She now seems to see math as something plural, as having "lots of different kinds. . . ."

**Figure 5–5.** Marnie's drawing (prepanel)

**Figure 5–6.** Marnie's drawing (postpanel)

# Final Thoughts

We believe that the mathematicians' panel exhibited great strengths in combating many of the stereotypical images that plague the field of mathematics, because it showed, in the words of one student, the "human face of mathematicians" and because it enabled students to know what mathematicians do. The panel also appears to have encouraged students to view mathematics as a study beyond mere calculation and number. The mathematicians themselves participated enthusiastically, happy to aid the mission of changing currently held stereotypes. We believe that such an intervention is easily replicated and can serve as a means for sensitizing students and perhaps even their teachers to the images they hold of mathematicians and mathematics. It thus seems hopeful that a much closer link between students and mathematicians can be forged. This bodes well for the future of mathematics and for a more enlightened general public.

# Works Cited

Emmer, M. 1990. "Mathematics and the Media." In *The Popularization of Mathematics*, eds. A. G. Howson and J.-P. Kahane. ICMI Study Series, 89-102. Cambridge: University Press.

Henrion, C. 1997. *Women in Mathematics: The Addition of Difference.* Bloomington and Indianapolis: Indiana University Press.

Howson, A. G., and J. P. Kahane. 1990. "A Study Overview." In *The Popularization of Mathematics.* ICMI Study Series, 1-37. Cambridge: University Press.

Picker, S. H., and J. S. Berry. 2000. "Investigating Pupils' Images of Mathematicians." *Educational Studies in Mathematics* 43 (1): 65–94.

———. 2001. "Your Students' Images of Mathematicians and Mathematics." *Mathematics Teaching in the Middle School* 7 (4): 202–209.

# Bridge

Susan Picker and her coauthor, John S. Berry, a distinguished mathematician from the Centre for Teaching Mathematics at The University of Plymouth in the U.K., have effectively joined the main issue of this book. Mathematics is no longer the "other." It has a human face. Mathematicians have bodies. They might do anything and everything: fold paper, play games, study the international patterns of disease. They laugh, wear cowboy hats, play music. They are themselves students who learn from one another and from their own students. It's a social process, just as Ian claimed.

We go next, then, in the peculiarly symmetrical logic of this book, to David Hardy, whose visceral and humane understanding of reality dictates that any problem worth solving is a worldly one. His chapter introduces us to the middle ground between math and the humanities.

# 6

## Life and Math at Imaginary High

### David Hardy

Introducing math into a humanities curriculum or humanities into a math curriculum in high school—why is this so hard? English teachers have little congress with math instructors, who themselves are not often found consulting with the social studies department. There must be reasons for this beyond the fact that teachers have their own tests to prepare and their own curricula to cover. Perhaps the problem is built into the inherent structures of the disciplines.

Math versus the humanities. To get a grip on this problem it helps me to think about Mr. Snickersly, the head of the math department in an imaginary high school I carry around in a kind of parallel universe in my mind. Mr. Snickersly has been at Imaginary High for twenty years. During this time his hair has become thinner while his middle has grown thicker, but his love of math is undiminished.

Mr. Snickersly loves all of it, the whole system: axioms, theorems, and proofs. He sees this system as a tree that grows from a seed that consists of a very few fundamental assumptions. Every leaf, every twig, and every branch of mathematics from algebra to string theory is a result of those first principles. No unproven speculation is allowed to graft itself onto the trunk. Fermat's excellent theorem remained uncontradicted but also unproved for 300 years. The fact that there had never been a counterexample even after three centuries of diligent search is no help when it is proof that is required. It comforts Mr. Snickersly to think that there are colleagues of his whose job is to rigorously check and double-check the credentials of any candidate for inclusion into the mathematical family tree. That the physical universe should occasionally provide evidence of the truth of a particular branch of the mathematical family interests Mr. Snickerlsy not at all. It is the abstract purity that Mr. Snickersly admires. He loves the tree for what it is, not for what it does.

For the rest of us math is a different kind of beast—one that is intimately connected to the manifest world. To us it is evident that the very idea of number is a result of the fact that the human mind cannot cope with the sensory world without separating it into a collection of entities (as opposed to an undifferentiated mush of being) and that it is a pragmatic necessity that these entities be categorized and counted. In other words, math exists because we need it. For us it makes a difference whether it is eleven dollars or eleven doughnuts. The fact that this number is prime interests us only when we must split the poker pot or doughnut box among fewer than eleven friends. Even Mr. Snickersly must count his change when he picks up his white shirts from the dry cleaner.

Although the general approach of the academic mathematician to teach the system, not the tool, seems natural to a math lover, it may not seem so to other types of students. Certain pragmatic individuals find their understanding of fractions on a ruler to be incomplete until they have been given something to measure. Others may go so far as to require an actual reason to measure, as they might in the disciplines outside of high school that regularly use math as a tool. Obviously, some disciplines in high school use math as a tool as well, usually ones in the science and technology wings of the institution. One of the classes I teach, however, Principles of Engineering, is in some ways even better suited to the examination of the particular uses and limits of math when used as a way to model the world outside of Imaginary High.

As the name suggests, Principles of Engineering is a course for students who may take engineering in college. Such students tend to be relatively ambitious and high functioning. It is often remarked that their math skills tend to be developed far beyond those of an average grown-up, whereas their aesthetic and moral capabilities seem much less so. This is one of the things that Principles of Engineering is designed to address. In fact, except for a similar college course, it may be the only time in a student's engineering education that humanistic concepts are mentioned.

One of the activities in my Principles of Engineering class is to design a bridge on the computer using a program called West Point Bridge Designer (available free on the Internet at *http://bridgecontest.usma.edu/download.htm*). The task is to design the least expensive bridge that will support a given load.

Students' first efforts usually result in a message that reads "Your structural model is unstable. The load test cannot be performed." Once an acceptable design is in place the computer will run the load test, which is represented by a nifty animated panel truck that attempts passage across the span. Usually the bridge members bend, snap, and dump the truck into the river below. Students love this, but they eventually move on to build bridges that hold the weight of the truck.

The second phase of the task is to redesign the bridge to cost as little as possible while still keeping the truck dry. A box at the bottom of the page keeps a running account of the cost of the student's design. As each structural member is added the price increases at a rate that depends on the size of the member. Students can call up a page that gives a detailed analysis of the load on each member and its designed capabilities. With this as an aid the students can refine their bridges so that no member is overbuilt for the job. The bridges become increasingly elegant as the process continues. No student yet has designed a bridge that was both ugly and cheap.

Obviously, this particular design process depends on the prodigious number-crunching capabilities of the personal computer. It is important that the students understand that the computer model is simply the math disguised and that math itself can be used to model the world of trucks and bridges.

If an aesthetic appreciation—perhaps appreciation of form is a better way to put it—can be abetted by well-crunched numbers, the same cannot be said for an ethical one. A traditional function of an introductory engineering course at both the high school and college levels is to imbue the students with a sense of the moral responsibility of engineers. Students are taught that their first allegiance is to the public. Most seem to grasp this principle. The difficulty comes as interests overlap and outcomes become murky. Sometimes math helps clarify the problem. Here is a question from my final exam:

> You are an engineer in charge of testing fasteners for a small company. You sample five of the latest batch of bolts (10,000) that are to be shipped to a major client. All bolts pass the shear strength test except one, which fails a few pounds below the shear strength required for bolts of that type. You run additional tests to determine the cause of the failure but find nothing. The boss tells you to go ahead and ship that batch. What do you do? Do you have any responsibility in this situation? Why?

Some students ship the bolts and some don't. Some take the trouble to figure out that if five bolts are a representative sample, then 2000 bolts could be defective. These students usually don't ship the batch. Some students will suggest that there be more mathematical clarity introduced by testing a larger sample. When pushed, all students will agree on the math that is inherent in the question. The same is not true of the answer, because it is a moral judgment—a value judgment—and math is value neutral. The math is the same whether the fasteners to be shipped are bolts or bone screws or shoelaces.

Here is another example that comes straight from a college engineering textbook:

### Case One: The Overcrowded Lifeboat

The time is one hundred years ago, and a sailing ship carrying passengers and cargo to Europe hit an iceberg and sank. Fifty people survived and tried to crowd into a lifeboat intended for twenty. A storm threatened, and the captain decided that some of the people would have to be cast from the boat and left to drown. The captain judged that only the strongest should remain on the boat, as they were needed to row. After days of rowing, the survivors were rescued, and the captain was put on trial for his actions. He justified his action because those who drowned would have done so anyway, and doing nothing and allowing all to perish was a worse course of action. Put yourself in the captain's situation. What would you do? Why? Is it different if you are a member of a jury judging the action? (Burghardt 1995, 93)

This has all the earmarks of a math problem, but the math is of little use. It may help us determine a risk/benefit equation based on past experience. Perhaps there is a 40 percent better chance of survival after five days in a lifeboat that is 20 percent underloaded than in a lifeboat that is 30 percent overloaded. You can then figure the increased odds of survival of the remaining passengers against the almost certain death of the ones you had to toss overboard in order to get these improved odds, and so on. Although this risk/benefit calculation is crucial to insurance actuaries it is of little help to cancer patients or to lifeboat passengers. In my class there is always a group of students who take the hard-edged accounting office view of the rowboat problem and are willing to chuck anybody overboard under the Social Darwinist credo "only the fit survive." These students will insist on this path up to the moment you fill the boat with their mothers, girlfriends, or baby brothers.

It is all about how you set up the problem. For structures like bridges we are often calculating the load that may be applied to that structure and then designing that structure to withstand that load plus a bit extra just in case. The "just in case" is called a *factor of safety* and is a hedge against the variables that we cannot perfectly anticipate. The percentage of extra strength varies with the value judgments that are made about the use of the structure. The factor of safety applied to a tightrope cable will thus be quite different from that applied to a shoelace, even though the math is essentially the same.

One of the things we try to do in Principles of Engineering is keep the math connected to the artifact. Later, in engineering school, as the student works with increasingly sophisticated mathematical modeling, it will be easy to forget what the math is intended to accomplish. Engineering schools are beginning to take this into account by offering courses that give students a taste of real-world engineering earlier in their college careers.

Math exists because it is crucial to know whether the four of us are facing three tigers or one tiger. Indeed, we may have to communicate that information among ourselves. We may even need to know how many tigers are left in the immediate bush if we are lucky enough to kill one. Is there a mathematical model that will help us choose which tiger to attack? Maybe weight would help. Maybe not. Maybe we have to redefine the problem based on new information that one of the tigers is a mother intent on protecting her two cubs. Does it matter whether she is protecting two cubs or three cubs? What do the tiger's value judgments have to do with the math of this problem? What about us? Are we ethically required to protect one another? If so, how does this affect the equation? Perhaps a statistical inquiry into the number of persons who have survived tiger attacks would be helpful. Perhaps a risk/benefit assessment as applied to the four of us would be useful. Perhaps, since it is our own personal butt at stake, it would not.

Using math to model our universe is a natural human activity. In many cases the more sophisticated the model, the farther it is from the seat of our pants, making it more difficult to assess its accuracy. Its accuracy in relation to itself, Mr. Snickersly's ideal system, may never be in question, but the accuracy in predicting the permutations of the universe, relying as it does on the intelligent application of the system, may be in doubt. Although mathematical proof will sanctify the appropriateness of a new addition to the great tree of mathematical life, the only thing that ratifies our applications of these notions as a tool are the results among the tigers, bridge crossers, and lifeboat passengers of the actual world.

The engineers (and scientists) are the ones who live in the space between this world and the math built on centuries of painstaking proof. It is they who scurry between the big tree of mathematical knowledge and the physical world seeking correlation. The effects of their efforts are too often ignored by mathematicians and dentists and social studies teachers until a bridge collapses or a plane falls down. It is only then that some of us begin to understand the important work that gets done between the perceivable human world and the abstract mathematical universe.

# Work Cited

Burghardt, M. D. 1995. *Introduction to the Engineering Profession*. New York: Harper Collins.

# Bridge

David Hardy happens to be a close friend from college days. One of the things he does when he is not teaching is to build finely crafted wooden boats. In the 1980s he decided to build one from a design by the turn-of-the-century boat builder Albert Strange. He lofted it in a tiny apartment in the depths of the Lower East Side a few years before he moved upstate to Cold Spring. (*Lofting*, in which you map out your boat's plan life-size in two points of view, is the first step in turning the pure math of design into the messy stuff of real-life wood, glue, and varnish.) After lofting it in his apartment, he constructed the boat in a basement on Christopher Street. Occasionally I would drop by to see how it was going. He was forever gauging by eye whether the curves of his planks were "fair," that is, *looking* mathematically precise as they curved and twisted along the contours of the frame. David embodies more than any individual I know the real-life application of mathematics. His convictions are in his bones. It's a good thing they run so deep because it took a good amount of belief to get the boat out of the basement when it was finished. Picture getting a model ship back out of a bottle.

This famous passage from "Mathematics and the Metaphysicians," by Bertrand Russell, defines further what may be meant by the phrase "pure mathematics":

> Pure mathematics consists entirely of assertions to the effect that, if such and such a proposition is true of *anything*, then such and such another proposition is true of that thing. It is essential not to discuss whether the first proposition is really true, and not to mention what the anything is, of which it is supposed to be true. Both these points would belong to applied mathematics. We start, in pure mathematics, from certain rules of inference, by which we can infer that *if* one proposition is true, then so is some other proposition. These rules of inference constitute the major part of the principles of formal logic. We then take any hypothesis that seems amusing, and deduce its consequences. *If* our hypothesis is about *anything*, and not about some one or more particular things, then our deductions constitute mathematics. Thus mathematics may be defined as the subject in which we never know what we are talking about, nor whether what we are saying is true. People who have been puzzled by the beginnings of mathematics will, I hope, find comfort in this definition, and will probably agree that it is accurate. (Russell, 1956, 1576–1577)

The issue of whether it is better to teach pure or applied mathematics in the secondary school classroom has been explored in detail by John S. Berry, the co-author of the chapter "The Human Face of Mathematics: Changing

Misconceptions." It was my pleasure to be able to attend a workshop on the subject conducted by him in April 2001. In his dryly humorous way John led us through an experience that revealed with the deepest understanding the difference between math *modeling* and a math *investigation.*

He explained that a problem-solving activity to develop *modeling* skills involves the following features:

- The problem is described in terms of a realistic situation.
- The problem is real or of interest to the pupil.
- The problem has a variety of approaches and uses a range of mathematical ideas.
- It can last for a few lessons to many weeks, developing "staying power."
- The problem is open-ended and contains many questions.
- The activity ends with an outcome in the form of a poster or report.

These criteria describe what David Hardy asks his students to do in his engineering class. On the other hand, we have John's explanation of a math *investigation:*

- Mathematics investigations are pure mathematics puzzle-type problems.
- They can provide an enjoyable experience of "playing" with mathematics.
- They involve an exploration of mathematical problems that does not necessarily involve real application.
- They can be used as an approach to learning and teaching.

After explaining the fundamental differences between modeling and investigations, John asked us to work in pairs on our choice of a modeling activity or an investigation. I chose an investigation problem, in this case one called "Leapfrogs." ("Mathematicians' frogs are not the same as real frogs," John noted.) The problem is descibed this way:

> Six people are divided into two teams of three. They are sitting on chairs with an empty space between the teams. The people are allowed to move in two ways: (1) they can *slide* to an empty seat or (2) they can *"jump"* over one person into an empty seat. Find the minimum number of moves (slides and jumps) so that the teams have changed places. Investigate what happens if you increase the number of people.

My partner was a calculus teacher. At first I was daunted to be paired with someone at her skill level, but she didn't have any handy answers in her pocket any more than I did, so we attacked the problem together. We tried different patterns of leaping and sliding. I moved the "frogs" and she recorded the moves. We worked on different kinds of notation. We drew diagrams. She scribbled algebraic notation. For a while I found our setbacks frustrating. Then, I began to feel we were on the verge of a solution. "I think there's a pattern here, or some kind of rule," I said. John happened to be walking by and raised his eyebrows encouragingly, which whetted my appetite to continue. In a classroom setting we would have had time to find our solution, but John had to stop us to move the workshop forward. He lined us up with other teachers in front of the group and moved our bodies like chess pieces to demonstrate the very pattern that we had been on the verge of finding, which, with Ss standing for slides and Hs standing for hops, looks like this:

S-H-S-H-H-S-H-H-H-S-H-H-S-H-S

Elegant. Perfectly symmetrical, with three hops in the middle. Was that significant? Indeed. John informed us that there will always be as many hops in the center as there are frogs on a side: 108 or 3001 or 1,000,007 or 2 frogs on the side: You'll have that many hops in the center.

SLIDE

HOP

SLIDE

HOP HOP

SLIDE

HOP HOP HOP

SLIDE

HOP HOP

SLIDE

HOP

SLIDE

I can't help playing with the pattern even now. The choreography for a frog fandango. A concrete poem. A calligram:

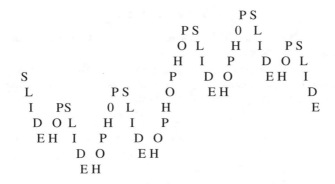

Your eye starts to do weird things, eh? Slip up sloppily, see dolphins, do lips, pop old hips, shop. And damned if it doesn't look like the formula for an interesting story plot! (See the Kurt Vonnegut example in the appendix for ideas on how to turn this pattern into a story.)

John asked us to take notes on our experience after working on our problems. I wrote: "manipulatives; record keeping; practice that develops intuition; patterns; rules; generalizations; interest in algebra *when it's derived*; not before the experience—when it serves a purpose."

One of the great lessons for me was the discovery of my own intuition within the process of the investigation. After my first thought that a pattern was embedded in the process, I doubted myself, only to have my intuition prove correct. We have already discussed how the *pleasure* of educating our intuition is what will sustain us all through the frustration of learning. Indeed, John told a story of how he first reacted to the kind of modeling exercise proposed in the ARISE curriculum (COMAP, Inc. 1998). "Here I was, a PhD in mathematics, throwing up my hands in frustration. Yet five minutes later I found the solution." John thinks that all math teachers should have this experience before they teach through modeling.

I have slipped inadvertently into a discussion of modeling from a discussion of investigations, but the point about intuition applies to both. In fact, John's contention is—and I agree—that investigations, with their emphasis on simplification, notation, and finding patterns and structure, make us better math *thinkers*, which can only make us better math *modelers*.

So what is wrong with David's Mr. Snickersly? The way I see it, he doesn't give students time to develop their intuition through exploration and discovery but expects them to understand merely through exercises. "The beauty is there, look at it, but don't touch," he would say. Constructivists like John Berry would say, "Can beauty be found there? Here are some tools. Dig in and see if you can find it."

Peter Dubno, in our next chapter, shows how to use mathematical tools in the context of interdisciplinary work.

# Works Cited

COMAP, Inc. 1998. *Mathematics: Modeling Our World.* Cincinnati: South-Western Educational Publishing.

Russell, B. 1956. "Mathematics and the Metaphysicians," In *The World of Mathematics: A Small Library of Mathematics from A'h-Mosé the Scribe to Albert Einstein*, ed. J. R. Newman. New York: Simon and Schuster.

# 7

## Math and the Westward Expansion

*How an Interdisciplinary Project Changed My Thinking*

### Peter Dubno, Jr.

At first it seemed foreign to me to think of mathematics and history being entwined in the same breath, let alone the same project. I had been teaching mathematics to middle school students for twenty-seven years and the only history I ever taught was how the king of England invented the unit of measurement called the foot, or how Pythagoras created his famous theorem. I was too focused on preparing my students for standardized exams to delve any deeper than that. But that was in the past. In my new school, the New York City Lab School for Collaborative Studies, where I began to work in 1997, I was encouraged to work with my grade colleagues as a team, and not to worry so much about state tests. It had been the experience of the school that this shift in emphasis resulted in the students doing just fine on them. The kids were happier learners.

My eighth-grade team meets once a week for two periods to plan projects and solve gradewide problems. It was at one of these meetings in 1999 that the English and history teachers asked me if I would be interested in participating in a project that they were very excited about. They were feeding off each other about how they could bring the westward expansion of the United States alive for our students. It was fascinating to watch this creativity bubble out before me, and none of this was going on at the behest or oversight of an administrator of any kind. "Hey, I can be as creative as easily as the next person!" I thought to myself. Despite the fact that spending time, perhaps a week, to do an integrated subject project was raising my red flag of caution, I decided to take a risk and change my usual thinking about the pace of my curriculum. "Sure, I can bring some mathematics into the project," I bravely told my colleagues.

Our plan allowed me three weeks to come up with some ideas on how math might be a part of the westward expansion unit. Thinking about it, I realized I had always enjoyed learning about the states in my own schooling, and I had traveled to many states in my life. I had an appreciation of my own country

and its idea of states' rights. It had always impressed me that each state in
the union is responsible for making up and deciding certain laws, procedures,
and bureaucracies. Each state has its own educational system—fifty individual
curricula, and school systems. It's almost like having fifty different countries.
Fifty different departments of motor vehicles, legislatures, senate buildings,
and so forth. Every state is unique, and yet we all belong to the same country.
There's beauty in this, but it is also flawed. On the international scale, I recalled
that at one point I had discovered that Japan is roughly the size of Montana,
which goes to show how large Montana is, or, conversely, how small Japan
is. Japan is a whole country, a whole culture. Such contrasts are amazing to
me. I asked myself, while looking at a map of the United States, "Why do so
many western states have squared borders while there are so many irregular
ones in the east? Why are most state capitals located roughly in the center of
the states? Why does Nebraska start precisely there, and Kansas stop? Could
this have something to do with latitude and longitude? What's the scale? Isn't
that math?" The more I looked at a map of the United States, the more math I
found.

Talking with my own eighth graders, I found they knew very little about
their country. Many had no idea of the name of their own state capital, let alone
surrounding ones. One of my very brightest math students confessed he had
never heard of Iowa when I asked him to locate it after math class. Another
student had to ask if Massachusetts was "big or small" after another student
challenged her to find it on the map. "Boy," I thought, "do our kids need some
geographical help." Could math help them learn this?

After talking with my colleagues and my staff developer, I put down on
paper a series of math-related investigations to help my students learn math in
the context of the westward expansion. (See the following project description.)
I assigned each student a state in the expansion. The original thirteen were
excluded. The students then had ten minutes to trade states with one another
if they wished. I had to laugh when students pleaded, "I'll trade you Texas for
Colorado. Colorado is so easy because it's square and Texas is so big." (Both
ideas are immaterial to the project.)

After these negotiations were finalized, I introduced the students to various resources. To my surprise, most of them had never heard of an almanac
or the U.S. census and were afraid to turn the pages of a real encyclopedia.
They wanted to use the Internet to get *all* the required information. This wor-
ried me, as not everything is accessible on the Internet, and the information
that can be found may well be obscure or inconclusive. They had to do proportional drawings of their states (no tracing permitted); research the borders, find
pertinent dates; measure routes of travel for the past and present (provided to
me by the history teacher, Kerry McKibbin), including ratios; name important
cities of past and present; name the state capitals of past and present, make
charts of population growth from the date of statehood to the present; write

linear equations to represent the growth; and find the population density of past and present times. This was a lot. How was I going to apportion it in my curriculum?

I decided to take five math periods over a ten-day span to accomplish my part of the project. During this time, I helped with the gathering of statistics and gave the students library and computer room passes. To help them measure routes of travel I handed out string to find the lengths of rivers on a large map, and referred them to the scale found on the legend. I also taught scale drawing.

In the end, this wasn't a great amount of math time to devote to the project, and my students learned an awful lot of relevant math, not to mention much about their own country. For the first time, many took a close look at a U.S. map. They saw how the states related to one another. They discovered what natural resources each of their states held. They learned there were other important rivers besides the mighty Mississippi. They looked at published U.S. census statistics and were able to see trends in them.

The students were permitted to present their findings in the form of reports or displays on oak tag, construction paper, foam board, videotape, or any combination of these modes. They were all very excited about their work and came back with fascinating historical stories of the events that determined the borders of states. North Carolina at one time included Tennessee but had to be divided because citizens couldn't cross the Appalachian Mountains to get to the capital. Toledo is situated on the border between Michigan and Ohio. Both states wanted to claim this important commercial city. The citizens of Toledo took up arms and were about to fight over which state they belonged to. Hearing of this, Congress settled the dispute by assigning Toledo to Ohio and appeasing Michigan with the Upper Peninsula.

It was astonishing for me to witness firsthand what can happen when humanities and math people put their heads together. It creates a symbiotic way of thinking that produces something greater than anything that can be achieved by one subject alone. It freed my thinking. I realized you can find the math in many diverse directions and situations. Of course there are limits, as well. Another math teacher and I conceived a plan to make a giant puzzle of the states for our project room floor, but the states had different levels of detail when we cut them out, and solving this technical problem was too onerous for the time we had. The payoff in learning simply wouldn't be great enough. The lesson here is to use good judgment.

Our collaboration seemed to help the other teachers too. I taught Kerry McKibbin how to make scale drawings and how to use string to measure the lengths of rivers. Both of these techniques have found a place in her history lessons. She has also become more comfortable with the use of graphs and statistics, which she uses in her students' study of the Great Depression and a project on imperialism. The students see teachers teaching teachers and learn an important lesson from this: that we can never stop learning from one another.

I certainly learned a lot from them after their research. It really fleshed out my own knowledge in an area of high interest to me.

The Westward Expansion Project significantly deepened my own teaching, too. It was interesting the first year, but the more I did it, the more it enriched my curriculum. I noticed, for instance, that population growth in the states tends to be linear. My students happen to study linear relationships later in the year. I now send them back to their state data to find lines of best fit. The circling back and the links are useful in reiterating important concepts.

Since the westward expansion project I have done other interdisciplinary projects such as the bridge project and the audiotape project described below. The work my students did on their states helped greatly with an audiotape project for a theatrical production of *To Kill a Mockingbird*. The students took the statistics and trends from their studies and used them to analyze the social life in the country during the period in question, from 1870 to 1930. Entwining math and the humanities no longer seems so alien to me now, especially when I repeatedly see how the students began to sense the connectedness of their studies. I am convinced now that the essence of teaching is not just marching along to the tune of a textbook but knowing how deep to go at various times.

Here is the project description as I gave it to my students.

KNOWING A STATE
FORMED BY WESTWARD EXPANSION
Westward Expansion Project
Math Component

Due in ten school days after assigned.

You have been studying America's "westward expansion" in history class. This is your opportunity to really research some of the history, geopolitical decisions, travel routes, and population growth of your assigned state. You will have to include the following requirements of the project. Of course you may add more, but don't substitute "more" for the requirements. A list of resources is given on the back of the sheet.

### Requirements

1.  Proportional Drawing: Draw a map of your state to scale on graph or grid paper. (No tracing, with this exception: you may present your whole project in the correct shape of your state.)

2.  Map Research: Find the four, five, or six borders of your state. How was each of these borders decided? Are they latitude or longitude lines? Exactly

what latitude or longitude? Are they natural borders? DO NOT say "They were the 'ends' of the state" or "My state ends because that's where Kansas starts." Find out why Kansas' border ended and your state started.

3.  Birth Date: Exactly when was your state born (the day it became a state)?

4.  How Long—How Wide?: What were the major "routes of travel" for your state when it was born? Did they have names? How long, in miles, were they? How long, in time, did it take to travel these routes?

5.  Ratio: What are today's "routes of travel" in your state? Do they trace any of the original routes of travel? Why? How long, in miles, are they? You can find this information on a road map of your state. Find the ratio of any one or two original routes' length and today's replacement route. Reduce this ratio to lowest terms or a good approximation.

6.  Capital. What was the capital of your state when it was born? What is it today? If it changed, why did it change?

7.  Before and Now: List the three or four most important cities or towns of your state when it was born. Are they still the same three or four most important cities today? Did any of them vanish over time? Why?

8.  Rate of Growth: Find the population at your state's birth and every ten to thirty years hence (up to today). Make a chart or graph of your findings. Can you say that this "rate of growth" between each of these ten- to thirty-year periods is linear or some other pattern? If it is linear, or close to linear, try to write an equation that models this "rate of growth."

9.  Density: The population density of a state, an average, is a value that represents how the number of people who live in the state fit into all the land in the state. Thus

*Density = (Total population) ÷ (Total square miles in a state)*

The higher the density number is, the denser the state population. Find the density of your state when it was born, and the density today. What did you notice when you compared these two densities?

### Sources of Information

1.  Any encyclopedia
2.  The *World Almanac*—usually valuable
3.  Any library, especially ours at NYC Lab School. It has books on all states.
4.  State chamber of commerce
5.  History or social studies textbooks
6.  U.S. census
7.  Road maps or atlas

8.   On-line services—In the past this has been just a so-so source for this info.

9.   Your parents and teachers

How to present your project:
You can present your work on oak tag, poster board, foam board, large construction paper, videotape, report, or using a combination of modes. Make sure you have *all* the requirements. *The requirements are what you will be graded on.* If you get stuck on something, please see your teacher as soon as possible. This should be a fun project for you. If it isn't, then you're doing something wrong.

Here is an example of a student's work on the project (unedited):

Devin Flaherty 802
12/16

## MATH STATE PROJECT: TEXAS

For the math state project I was given Texas. It was a pretty hard state to map because it's a shoreline state with really curvy borders. I researched my state and the reasons the borders were formed. I researched past and present populations, the state's birth date, past and present routes of travel, density and more. Also I made a proportional drawing of the state. Below is the information that I found.

Texas became a state on Dec. 29, 1845, shortly after the Mexican War. All of Texas' nonnatural borders were decided when the Mexican War was lost by Mexico to the U.S. and the two countries made a treaty in which Mexico turned over all of the American southwest including present day California, New Mexico, and Texas. The natural borders are the Rio Grande on the southwest border which runs diagonal between the 97° and 107° latitude, the Red River on the northeast border which runs on the 34° parallel, the Sabine River on the east border which runs latitudinal on the 94°, and the Gulf of Mexico on the southeast border which runs diagonal between 94° and 97° latitudinal. None of these borders have changed.

In my research I found the population census for Texas every ten years from 1850–1970. I found that the population growth every ten years didn't even come close to a pattern. In some decades, the population went up 25% while in other decades it went up as much as 60%. I believe that this drastic difference is due to the years when more people were moving westward and years when people didn't. The population never went down but it always kept growing at an extraordinary rate. Only for a few decades between 1880 and 1910 was the population growth even close to linear. Though I did find that every 30 years the population seems to roughly double. I made a linear equation but due to the very varied rate of growth it is quite rough. Here are some examples of population rate:

*Linear equation:* Present population = pop. 30 years ago × 2

This linear equation would only work if the rate was a bit less varied. Below in the chart I show the percentages of growth every 30 years and below that, a graph for the *actual* population every ten years. I couldn't put the two groups of data into one graph because the data for every thirty years wouldn't fit with the graph I needed to make for every 10 years.

Texas is the second largest state in size and in the top five in population. It's capital at birth and today is Austin. Austin itself has a population of over 520,000. Its largest city, Houston, reaches over 3,000,000. Its total population as of 1997 was 19,439,337. Its area is 267,277 square miles. With a population of 19,439,337 its density is 72.7 people per square mile as of 1997. The density today, based on the 2000 pop. Estimate of 22,362,000, is about 83.6 people per square mile. Back when Texas was born a state in 1845, with a population of only 213,000, the density was 1.3 people per square mile. That means the density today is 64 times greater than the density when Texas was born. This would create a ratio of 64:1.

Both now and when Texas was born there were major routes of travel. An interesting fact that I learned was that a travel route from the 1850's called the Arkansas Route travels almost exactly where Interstate 66 runs today. The difference between them is that Route 66 covers over 4000 miles from San Francisco all the way through the U.S. up to Montreal. The Arkansas Route ran from El Paso presently in New Mexico up through Texas to St. Louis for only about 1000 miles. The ratio here being approximately 4:1. Many routes of travel ran through or began at St. Louis in 1800's because it was, as it is today, a major city. There was also the Butterfield Overland Mail which also ran through Texas and stretched approximately 2200 miles from San Francisco to St. Louis. Both people and trade items started from St. Louis and ran west all the way to San Francisco on some routes.

Today, Texas has several major cities; all of which you have probably heard of. There is, of course, Austin, the capital, Houston, Dallas, San Antonio, and Galveston. Back when Texas was started, its major cities were Galveston, Austin and San Antonio. I researched it and there were no major cities when Texas was born that have vanished over time.

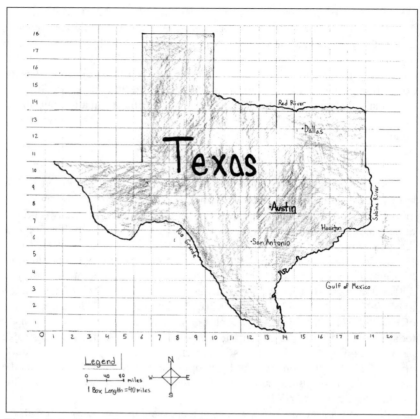

**Figure 7–1.** Proportional drawing of Texas

   Along with this report I have included my proportional drawing of Texas.
(see Figure 7–1). I increased the size of the drawing by increasing the size of
the grid boxes. I doubled the sides of the grid box therefore the original state
is 1/4 of the area of the enlarged drawing of Texas. But when the sides double
the area is four times bigger.

### Data Table for Every 30 Years

1850—213,000 = 13% (only five years after birth. Not a good
example of rate)

1880—1,592,000 = 40% of 1910

1910—3,897,000 = 60% of 1940

1940—6,415,000 = 60% of 1970

1970—11,197,000 = 50% of 2000 (estimated)

2000 (Estimated)—22,362,000

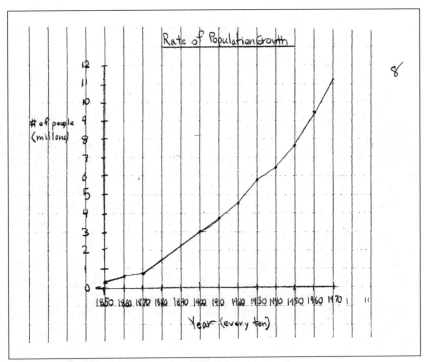

**Figure 7–2.** Graph of Population Growth 1850–1970.

As you can see from the graph of population growth (Figure 7–2), this data is far from linear or a straight line. Though for thirty years between 1880 and 1910 the equation is indeed very close to a straight line. But the rest of the graph is pretty curvy. On the data table above the graph (every 30 years), the population growth is almost linear. It's rough but the doubling pattern is noticeable.

# Bridge

Peter happens to be a trout fisherman, like Ian. He happens to be a handy guy to have around, ready to clean the air conditioner filter in your public school classroom that hasn't been touched for who knows how many years. He happens to be generous with his overhead projector, ready to loan it at any moment. He happens to be smart in a humble way that can enlighten you without making you feel stupid. (See my remarks on the book *The Joy of π* in the annotated bibliography to be found on the internet at *http://www.heinemann.com/shared/onlineresources/E00245/bibliography.pdf* for a conversation that illustrates this.) He happens to be, as far as I can tell, above all meanness and hypocrisy. But what makes him a collaborator is his ability to trust, once the trust has been earned.

To my way of thinking, if you are able to trust, you can collaborate on interdisciplinary projects. You can trust not only your partners, as Peter did, but you can trust yourself. "Our plan allowed me three weeks to come up with some ideas on how math might be a part of westward expansion. Thinking about it, I realized I had always enjoyed learning about the states in my own schooling. . . ." He trusted his own interests and his instinct that those interests would be relevant to the students because he was certain of the revelation in them. He knew the inquiry would have substance. He didn't force his interests on them, though. He assessed them first: "Talking to my own eighth-graders, I found they knew very little about their country. . . ." He consulted the expertise of others as well as that of his colleagues. This preparation made him worthy of the students' trust.

Trust, then, becomes a critical element in constructing a "continuum of experience" for the students—from subject to subject, but also through time. Weeks after the end of the Westward Expansion Project, the students worked on the other interdisciplinary project that Peter mentioned: The Mockingbird Monologues. In Erick Gordon's English class the students study Harper Lee's *To Kill a Mockingbird* and write monologues based on its characters. In their history class, with their teacher Kerry McKibbin, they study the history of racism after the Civil War and how it contributed to the events described in the novel. The climax of the project is a full-fledged performance of the monologues in a professional theater space. Erick describes the process thoroughly in his essay "Occupying Spaces: The Mockingbird Monologues" (Vinz et al. 2000, 8–42). Now that Peter was a part of the collaborative mix on this grade-team, he was able to make the contribution he described, in which "students took the statistics and trends from their studies and used them to analyze the social life in the country during the period in question, from 1870–1930." These facts were a natural expression of the mathematics of social and economic history and a reiteration of the students' earlier learning. Here is a small sample from one class' script:

> In 1924 there were close to five million Ku Klux Klan members in America. 115,000 of them lived in Alabama... In the 1930s white people made up nearly ninety-three percent of New Mexico's population.... The 1920's were very prosperous times. From 1924 to 1929, some stocks rose fivefold in value.... On October 24, 1929, the stock market crashed. The value of all stocks fell by fourteen billion dollars.... Social Security was set up in 1935 to benefit those who were retired, unemployed, or disabled as a result of the depression.... In 1923 the movie *Birth of a Nation* was released. It told the false story of "black savages" chasing "white virgins."

As in the "living newspapers" produced by the Federal Theater Project of that same era, this part of the performance relied on what theater artists of the time called "the entertainment value of the fact."

Kerry reported that the students had by now assumed a palpable ownership of their assigned states. They were far more authoritative in other areas of American history about geographic borders, and more sophisticated about issues surrounding slavery. But the students weren't the only ones who had become more adept in their use of math in history. A few months later I was observing Kerry's class when she started the day's lesson with the following word problem:

> A farmer owns a square plot of land. He has five sons. The first son gets one quarter of the land. The other four sons must be given land the same size and shape. Draw how this might look.

After class I asked Kerry why she had given the students this problem. Someone had shown it to her in a workshop, and she found it intriguing. She wanted to see how it worked with the students. It was just an experiment. Later in the same class the students worked in pairs preparing for a discussion on the economics of the Great Depression by plotting coordinate graphs of the effect of unemployment on average personal income. She had begun to use other skills she had picked up from the collaboration as well, such as scaling, and the use of threads to measure meandering rivers. "Collaboration deepens the learning experience for everyone involved," she said. "And when teachers learn from teachers it models the process for students. When math people and humanities people put their heads together for project work, the different kinds of thinking are symbiotic and produce something far greater than anything that could be achieved in one subject alone."

Interdisciplinary project work has the potential to be the most powerful way to realize our theme. Planned and conducted effectively, it provides authentic purpose to learning, involves students of varying ability levels in meaningful work together, generates multiple forms of assessment, contextualizes the acquisition of knowledge and the development of skills.

I know the power of projects personally. I've been involved in many in my interdisciplinary career. Some of the most notable have been plays I collaborated on as a writer with actors, designers, dancers, and composers. The productions took me on a path to meet distinguished historians, producers, and theater artists who would not have had any reason to share their worlds with me otherwise. The experience I've had with projects gave me the confidence to take on this book. From the outset I knew that it would work best as a kind of collaboration. What will never appear in these pages is the tone of the conversations I had with the contributors and other colleagues who advised me on it. The experience of collaborating, of coming to understand new things more deeply than one could alone, is the true reward of working on it, not the bound volume itself, which will seem like a mere husk once it is done. (Though I hope the words create a kind of offstage conversation of some inspiration to readers.)

The students who participated in the Westward Expansion and Mockingbird Projects had similarly deep experiences. They may forget facts or details, but they will never forget the experience. They'll always see math, history, and English as inextricably entwined.

Our next chapter comprises a series of letters between sixth-grade humanities teacher Kay Rothman and this author. Kay, like Peter, works at the Lab School. An incident in her classroom involving Peter illustrates further the value of trust and is worth relating here as an introduction to the chapter. Kay's students had been studying *The Iliad*. They had drawn graphs rating the level of importance of the main events of the story. Listing them along the $x$-axis, they then rated them on a scale of one to ten along the $y$-axis. One group of students, for instance, had given Hecuba's dream about giving birth to live snakes a nine on their ten-point scale, then followed that with four points for the birth (and immediate death sentence) of the Prince of Troy. A line connects the two events. When Peter saw their graph posted on a bulletin board in the hallway, he observed that the students had used the wrong kind of graph for the information being conveyed. The connected dots on their coordinate graphs indicated a cause-and-effect relationship between the events. Although there indeed may *be* a cause-and-effect relationship between the events (and again I refer to the Kurt Vonnegut essay, with accompanying student samples, for demonstrating how a graph can depict this), the information presented was limited to the relative *importance* of the events. A more accurate kind of graph, therefore, would have been a bar graph, not a coordinate graph.

When Peter pointed this out to Kay, he said, "It's okay. They'll learn it in eighth-grade." But this was unacceptable to Kay. Some of his eighth-grade math students had to come into her class immediately to teach her students the difference!

The anecdote conveys several points pertinent to this book. It is further proof of Peter's ability to inform colleagues about mathematics without making them feel stupid; it is an illustration of Kay's courage in using math practice in her humanities classroom. And it says something about the culture of the school; their administrators must support teachers if they are going to collaborate and experiment. The directors of the Lab School, Sheila Breslaw and Rob Menken, provide for this support by including grade-team meetings in the teachers' schedules, by articulating their collaborative philosophy in the school's philosophy statement, and by signaling the importance of collaboration in the very name of the school (The New York City Lab School for Collaborative Studies). Most importantly, they support teachers by their belief in collaboration and collegiality. Signals are sent through even the most inconsequential-seeming decisions from their office. Teachers with collaborating instincts are hired. The directors themselves put aside what they are doing to listen when teachers drop by the office. Parents are informed that their children will be assessed as much for their willingness to collaborate with one another as for their performance as individuals. They would expect Kay to try using graphs in her study of literature, and for Peter to say something if he could help her sharpen her practice. The result in this case is that an ideal "teaching moment" occurred when Peter's eighth graders reiterated their own learning by advancing Kay's sixth graders' understanding of graphs.

Other examples of Kay's willingness to experiment with mathematical practices in her classroom include inviting students to create tables that show change in New York City's demography and having them generate graphic illustrations of large numbers to give them a sense of the scale of historic time. Her students thus feel confident to undertake their own historical math explorations. Catherine, Rachel, Cindy, and Angelica produced a brochure to teach Mesopotamian Astronomy and Mathematics to the rest of the class. They went into great depth of discussion about the Mesopotamian concept of zero, compared Mesopotamian measurements with ours, illustrated the way numbers appeared in cuneiform script, and discussed Mesopotamian cosmology, astrology, and calendar design. At the end of their presentation they gave the class a quiz (see Figure on next page).

In deciding to develop this book, I realized that I was going to be responsible for the careful use of the terms in its title if it was going to have any validity. What exactly did I mean by *teaching, depth, math*, and *humanities*? It is the purpose of the book itself to define the first two. As it turns out, Kay and her students helped me find definitions for the second two. Her students had researched the word *humanities* in groups and posted their results on the wall. Here are four of them, syntax intact:

Humanities is the topic of learning that investigates human creations like literature, language and philosophy. Humanities is also the study of human culture, history and prehistoric times and modern life. It is also the most important subject in school.

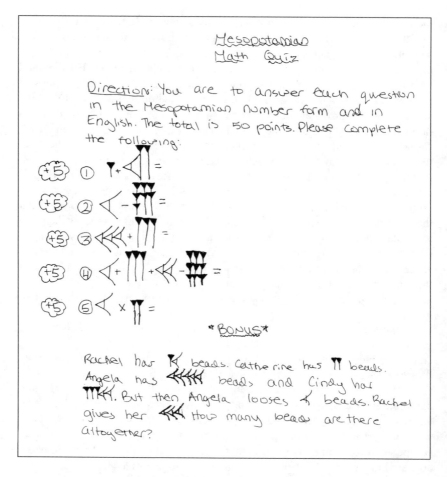

Humanities is the study of everything related to human culture. This includes the way we act, think and live. Our philosophy, our architecture, our literature, our art—all of this shows our different cultures. This is Humanities.

Humanities is the study of human development. How humans interact and the roles they play.

Humanities is the study of different civilizations and cultures told through sources such as literature. It is also a form of making connections with the world around us.

As for a working definition of *mathematics*, Kay pressed me to its definition in our chapter.

## Work Cited

Vinz, R. and E. Gordon, G. Hamilton, J. La Montague and B. Lundgren. 2000. "Occupying Spaces: The Mockingbird Monologues." In *Becoming (Other)wise: Enhancing Critical Reading Perspectives*. Portland, ME: Calendar Island.

# 8

## A Mathematical Correspondence Between Humanists

### Kay Rothman *and* Dale Worsley

9/29/00

Dear Dale,

It is 7 A.M. and I find that I must babble to you about one of those morning epiphanies that I claim to have. When I do math problems in my classroom, I invariably botch them. I talk them to the point of incoherence. When a math teacher I know tries to put his math into comprehensible English, he sounds crazed. In the book *The Teaching Gap* (Stigler and Hiebert 1999) the authors report that in Japan teachers don't introduce any outside analogies or things to make math "interesting" (basketball scores, stories). Instead they have math tell the story, because to them it is so exciting (go figure).

The other piece in that book is that the work on lesson design is done with other math teachers at the same level. Language teachers, art teachers, history teachers are not part of the mix. My contention is that they would muck it up! When I was doing art studies in museums, I felt that when people wrote about art it was certainly valuable on one level, but the mere translation of the visual to the analytical (the right to the left brain) took away, in some sense, from the experience of making or viewing art. My point: (I guess you were wondering when I would get to it) is that talking about math and writing about math is not the same as doing it.

The way in which it is not the same might have to do with the way our brains work. While some brilliant mathematicians might be able to write about it and come close, they are unusual in that they have the concomitant linguistic talent. So, Dale, what I am proposing to you (and you are the only one I think I would or could ever say this to) is that I think your philosophy on learning math might be wrong. As Emails carry a certain matter of fact tone and can be interpreted as harsh, I rest on our warm relationship to carry the tone.

Kay

9/29/00

Dear Kay,

How much are you prepared to correspond with me via Email about this? I don't want to even start unless you make a vow to keep going back and forth until we have an aesthetically pleasing series of letters. What do you say?

Dale

10/1/00

Dale,

I was out of town this weekend, but I can't think of anything I would rather do than write to you about this issue. You have my vow.

Kay

10/1/00

Kay,

Great. You're a real sport. I need to ask you some questions, coming from your epiphanic letter:

1.  What math problems are you doing in your classroom? (I love it that you would even think of it.) Are you talking about your generations/scale example?

2.  Exactly how do you botch them?

3.  Why are you doing them in the first place?

4.  I'd rather leave your math teacher friend out of it, if you don't mind, unless you need to use specific examples from him to make your points.

5.  I understand that in Japan they don't introduce outside analogies. (Though I do think I recall they sometimes introduce historical background.) You say that instead they "have math tell the story." And I couldn't agree more, by the way, that introducing so-called real-world connections, or outside analogies, is silly. But don't I score a point or two there, in the idea that math tells a *story*?

6.  Yes, the Japanese design lessons together as math-only teams. Excellent. And I couldn't agree more about language teachers, art teachers, history teachers planning math lessons. But understanding the principles behind each other's lessons—isn't that valid? Didn't you get a little something from reading about those Japanese math lessons? Maybe how they give the kids a problem and help them solve it, instead of showing the solution and having them memorize it?

7.  You say, "Talking about math and writing about math is not the same as doing it." I need to know here what you mean by "math." Do you mean

computation? Solving problems? Discerning patterns? Asking good questions? Memorizing notation? Predicting the stock market? What, exactly?

8. You say, "The way in which it is not the same might have to do with the way our brains work." There is a lot to support for you on that. (Much to be found in Stanislas Dehaene's book *The Number Sense* 1997.) There is also some interesting evidence to refute you—how the most basic number neurons and language neurons fire side by side.

9. You say my philosophy of learning math is wrong. What is my philosophy of learning math, exactly? (I won't put you on the spot. I'm actually simply working up a lot of talk around a thesis: "The teaching of math can make sense to students and its pleasures can be recovered if its isolation, its 'otherness' is understood and solved." Who knows if I can prove it . . . the talk is fun.)

So, let me know what you think when you have time. And read the Dehaene book? I'll try to run down a copy for you.

See you tomorrow.

Dale

10/9/00

Dear Dale,

After much discussion with my mathematically sophisticated nephew, I have come to understand my answer to question #7. What do I mean by math? I do not mean graphs and statistics, which are numbers used in order to show something else, numbers in the service of humanities, or, more usually, science. (I know that using statistics isn't math, because I excelled at it in graduate school.) I do mean asking good questions, but the questions asked are those with numbers. Solving problems that are posed numerically and answered numerically, is math.

The nearest I can come is that pure math involves an equation. Or that math, if it were a language, would have as its "sentence" a numerical equation. Its otherness is not to be denied, but needs to be accepted and celebrated (question #9.) It would be like saying you could learn and appreciate music by recognizing the music that was present in nature. One might be more attuned to the world of sound, but one's love of music and/or ability to play it would certainly not be enhanced (at least not that much). Music is learned and appreciated by listening to music and playing music. The joy that one can take in music is similar to the joy one can take in math.

That same math nephew (let's call him Charlie—that's his name) assures me that when he is trying to avoid real work or an unpleasant job, he takes out his calculus book and does a few problems. He reports that it works as an escape and, yes, a form of transcendence. This is, apparently possible. So how do we get people to find joy in math? Is it by understanding the ramifications of mathematics, its application in our life or its representation in the patterns of nature? No, I say. That is, respecting mathematics, something that I feel is an easy get. The only pathway might be through being so immersed in numbers, playing with them, somehow making them your own, that you experience a sort of love for them. Applying mathematics, or using mathematics is not going to get you there. It is backwards.

In my classroom, I *use* numbers: demographic tables that show change, or large numbers to indicate or illustrate time. I botch them up when I try to play with the numbers themselves, rather than use them. For instance, I might think that I can make a further point by having the kids combine different quantities. But in trying to explain how to do it, I realize I don't know how and get further from my point and don't know if it would have worked anyway. I talk about it some more, just enough so that I am aggravated and the children are confused. I know that isn't really an example, more like a description of the process.

All for now,

Kay

10/10/00

Kay,

Thanks for your amazing thoughts, to which I will reply when I have the time to give them a real answer. As for your nephew, you didn't tell me you were recruiting troops to ambush me. I might have to enlist someone into my camp for a counter attack.

Dale

11/12/00

Kay,

At last I can get back to you in our math correspondence. I'll go through your last letter a step at a time.

One of the things you say is that math is not graphs and statistics, "which are numbers used in order to show something else, numbers in the service of humanities, or, more usually, science." This is, actually, exactly one of the ways Morris Kline defines math. In his book *Mathematics in Western Culture* (1953), having described math as a creative activity, he asks what drives its practice, and points to three things: questions arising from social need, the need to rationally organize natural phenomena, and the purely intellectual challenge. The graphs and numbers clearly fall into Kline's first and second categories.

You go on to say that you know that using statistics isn't math, because you excelled at it in graduate school. This made me laugh out loud (or, as kids using the Internet write it: "LoL.") This, if we accept Kline's description of math, is in that same category of "questions arising from social need."

Finally, you do define your idea of math as "asking good questions, but the questions asked are those with numbers. Solving problems that are posed numerically and answered numerically, is math. The nearest I can come is that pure math involves an equation. Or that math, if it were a language would have as its 'sentence' a numerical equation." I'm actually going to support your point here. Math has created its own symbolic language, for efficiency of thought, but, as Kline puts it:

> Mathematics is more than a method, an art, a language. It is a body of knowledge with content that serves the physical and social scientist, the philosopher, the logician, and the artist; content that influences the doctrines of statesmen and theologians; content that satisfies the curiosity of the man who surveys the heavens and the man who muses on the sweetness of musical sounds; and content that has undeniably, if sometimes imperceptibly, shaped the course of modern history (9).

Almost in anticipation of Kline's point, you go on talking about the way one's ability to appreciate music might be enhanced by recognizing the music in nature, but one wouldn't learn how to play it better all that much. To learn and appreciate music, you contend, one must listen to it and play it. You call on your mercenary nephew again and talk about how he solves math problems for the pure joy of it. You think, apparently, that I disagree with you on this point. But I don't disagree at all. Practicing math is one of the things I wish everyone could enjoy. It seems to me that people like your nephew have somehow, through some miracle of their natures, survived the educational system's attempts to make math far more painful than it needs to be. Here is Kline one more time, if you'll allow me:

> [Math] has been a major cultural force in Western civilization.... [It] has determined the direction and content of much philosophic thought, has destroyed and rebuilt religious doctrines, has supplied substance to economic and political theories, has fashioned major painting, musical, architectural, and literary styles, has fathered our logic, and has furnished the best answers we have to fundamental questions about the nature of man and his universe. As the embodiment and most powerful advocate of the rational spirit, mathematics has invaded domains ruled by authority, custom and habit, and supplanted them as the arbiter of thought and action. Finally, as an incomparably fine human achievement mathematics offers satisfactions and aesthetic values at least equal to those offered by any other branch of our culture. Despite these by no means modest contributions to our life and thought, educated people almost universally reject mathematics as an intellectual interest. This attitude toward the subject is, in a sense, justified. School courses and books have presented "mathematics" as a series of apparently meaningless technical procedures. Such material is as representative of the subject as an account of the name,

position and function of every bone in the human skeleton is representative of
the living, thinking and emotional being called man (vii).

My contention, then, is that you have inadvertently, and unconsciously,
gotten trapped in the phenomenon of *school* math, the kind Kline describes as
"a series of apparently meaningless technical procedures."

You say, "So how do we get people to find joy in math? Is it by understanding
the ramifications of mathematics, its application in our life or its representation
in the patterns of nature? No, I say. That is, respecting mathematics, something
that I feel is an easy get." And I disagree with you there, because I think at least
some interest in math is generated by understanding its ramifications and its
representations in nature.

But you go on to say, "The only pathway might be through being so immersed in numbers, playing with them, somehow making them your own, that
you experience a sort of love for them." And I couldn't agree more with this. It
is certainly a variation on the theme of this book, and phrased better than I could
ever have done it. In fact, as far as I'm concerned, the idea that playing with
the thing to be learned, making it one's own until one loves it, goes straight to
my deepest beliefs about education. Schools need to be playgrounds, basically,
of the body and of the mind. It is from playing with numbers (along with every
other feature of math—numbers only being one) that we will love their nature
and come to understand them.

Finally in your letter you discuss using numbers effectively in your classroom but playing with them badly. Here I simply reiterate my point made earlier,
that you have, like so many of us, come to see the shimmeringly alive practice
of mathematics as singularly dead *school* math. As for the way you "use" numbers, I cheer—how many humanities teachers would be so ready to do this?
I wager that many would fear it. You're trying to "show change," illustrate a
thesis, and employing numbers to help you. Hooray!

One last point. When you say you botch it when you play with numbers, I
say of course you do. I would botch making sculpture at first, too, until I had
enough time and help in my play with the clay. I can hear you saying, "Yeah,
like I've got all day to go play with clay when I'm trying to teach eighty students
how to read and to write and to understand history into the bargain." I don't
have enough time to learn sculpture right now either. Maybe this is the line of
the Venn diagram I made in your class one day, where math class becomes the
part of the playground where you play with numbers—send the kids over there
to do it, or let them do it on their own like your nephew. You don't have to be
a math expert, I contend. All you have to do is let them know it's possible to
enjoy themselves and it's relevant not only to use math but play with it. (And
doesn't "play" mean being given permission to fail for a while?)

I honestly don't think we ultimately disagree with each other. I think it's a
matter of finding definitions that we agree on. I also think that you are a prime

case in what is driving me to put this book together: a person who has become alienated from math by education to the point the alienation seems natural. I await your reply. Notice that, in an effort to retaliate for your having hired mercenaries I have tried to overwhelm you with numbers—numbers of words, that is. Like Grant overwhelmed Lee with soldiers . . . except I think these words are fun to read. Do you agree?

Dale

8/18/01

Kay,

I know you're out of town and don't mean to ambush you when you come back, but guess what I've been doing? Working on our math book, which means, as you know, working on that bit of correspondence we had going. I've been looking at it and think it's worth finishing for the book. I don't think it would take much, actually. I've cut down one long segment of one of the Kline quotes. I think one more comment from you might be the way to end it. You would have the last word! Could I talk you into doing that?

As always,

Dale

8/21/01

Dear Dale,

The time lapse between now and when we started the correspondence has given me much time for idea percolation. I don't think we actually disagree on too many issues. I don't deny the power of using math as a tool, or the various ways in which math has impacted our lives (though I draw the line at Kline's assertion that math "has furnished the best answers we have to fundamental questions about the nature of man and his universe.")

I don't deny the power of using math in my classroom. I was trying to leap to the implications of teaching math. Maybe it is the distinction between "pure math" and applied math, BUT I still think that the distinction is real and that those who teach math should not mix the two up. I don't think that learning the "techniques" is the answer. Kline seems to be setting up a straw dog. Still, the best I can do is analogies like learning the language of music, not learning simply the scales and notation. Music exists outside of socially contrived language and, I still contend, so does math.

So, can I play with math and learn to enjoy it? Maybe. Will I? No. Why not, you ask? Again, an analogy: I played the piano as a child. I learned the scales; I learned the notes. I grew up on musicals—show tunes. I learned the words and sang and acted them out. I learned to play them on the piano. I had a great time playing the songs and belting out the lyrics. Music and lyrics went hand in hand for me. Later in life, I felt that as a well-rounded human being, I should try to learn about music—just music. I took a music appreciation course, learned the composers and what things were called. I listened to music for themes. Course completed, I listened to classical music for a while, but found myself relegating it to background music. Last Friday night, Ted, a friend of the notorious Charlie, was playing piano for us. I could recognize that he was brilliant. Afterwards he talked about how the piece—Beethoven—was so much fun and so accessible. The entire process thrilled him. I contend that music for Ted is a different animal, species, world, than it is for me. I can only use it to float a show tune. (Ted would be appalled.)

As for me and math in my classroom—I will continue to find ways to use it, to show kids things. I do think that math is a great and valuable tool. Quantification has the power of persuasion, as well as the ability to sell snake oil. But that's another issue.

Kay

## Works Cited

Dehaene, S. 1997. *The Number Sense: How the Mind Creates Mathematics*. New York: Oxford University Press.

Kline, M. 1953. *Mathematics in Western Culture*. New York: Oxford University Press.

Stigler, J. W., and J. Hiebert. 1999. *The Teaching Gap: Best Ideas from the World's Teachers for Improving Education in the Classroom*. New York: The Free Press.

# Bridge

So that's the kind of person Kay is. Someone who will have a "morning epiphany," tell a friend, and be able to run with it. Everyone should have a friend like Kay. Another humanities teacher unafraid to work with math to achieve greater depth of understanding is Avi Kline, who works at The Museum School, also in New York City's District Two, and housed, by coincidence, in the same building as the The Lab School.

# 9

## The Math of Art

### *Probing Design Principles of Classical Greece and Islam in the Classroom*

### Avram J. Kline

I teach English and Humanities to ninth and tenth graders at the New York City Museum School, a public school in Manhattan that partners museums with the classroom. The essential thing to know about our methodology is that it is highly visual; students regularly observe and analyze objects and art to understand themes in history. I see my role as prompting this inductive process through guided inquiry. In 1998 I worked with a museum educator, Rebecca Krucoff, and a global studies teacher, Jody Madell, to develop two museum-based units, one called "The Ideals of Classical Greece" and the other "Universal Religion in the Middle Ages." Each of us, through the lenses of our different disciplines, would research works of art that embodied the central ideas of the relevant cultures, come together to synthesize our findings, then teach the elements corresponding to our particular expertise.

To my own surprise and initial apprehension, I found myself bringing math ideas to the planning table rather than literary ones. I had come upon the famous golden rectangle in researching the Parthenon for the first unit, and Islamic tessellation in researching mosques, for the second. The logic, beauty, and utility of these two forms were irresistible. Here were two distinct concepts in *geometry* that would, if we could make sense of them, guide us through the principles behind Greek and Islamic art and architecture. In this chapter I will relate how my students and I explored these principles through trial and error, philosophical inquiry, and aesthetic intuition and how this foray into another discipline became an entertaining, delightfully tangential, even suspenseful teaching experience.

I opened my lesson on the Parthenon by projecting a slide of a distant view of the Acropolis and asking my students to ponder how the Athenians came to decide what to build on its summit. Pointing to the surviving edifice of the Parthenon, I asked rhetorically, "Was that large, columned building just thrown together, or was its design carefully planned to meet the values and desires of Athenian citizens?" I asked them to define the terms *building* and *architect*, then to brainstorm criteria for designing a successful public building. I wanted them to consider that the architect sets out to design a structure that is both functional and expressive of important ideas or cultural values. I also wanted them to sense the difficulty of squaring aesthetic values with pluralistic fashions. How does the architect consider what is best? A fruitfully ambiguous discussion on these questions led us to wonder, "*Is* there such a thing as universal taste?" I explained that I wasn't merely referring to the fleeting consensus behind trends. I wanted them to consider the possibility of testing aesthetic principles, that there might be certain shapes and proportions to which human beings are timelessly attracted, this being the virtue of the golden rectangle and the system of human proportion from which it derives.

I went on a tangent that I hoped would illuminate the matter. I had come upon an actual "rectangle survey," conducted by Gustav Fechner in the 1860s (Fechner 1860):

> Are certain shapes and forms simply more pleasing to the eye than others? Which rectangle do you find the most pleasing? Which do you find the least pleasing?

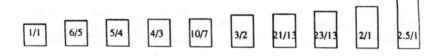

The majority of people Fechner polled evidently expressed a preference for one rectangle, its dimensions those of the very "golden rectangle" exalted by the Greeks. I tested it out on myself and came to the same conclusion. That a specific proportion of rectangle should *feel* superior and "perfect," is, I suppose, a primal gratification in geometry (one that precedes the intervention of numbers and logic). I predicted that if my students were to take the same survey, they would reach the same consensus, then marvel at how one proportion was simply more pleasing than the rest and we could dispatch the question about universal taste in visual art. Most of them found the survey unanswerable, though. And those who did manage to prefer one rectangle to the rest were not remotely unanimous in their choice. Was this indicative of how unaccustomed they were to critiquing the visual? Or were they practicing nonconformity for its own sake, their uncooperative response deriving from some culturally prevalent, postmodern relativism? I even considered the remote possibility that my students were simply less accustomed to golden rectangles in their ordinary lives than we have been in the past, compact disks having replaced rectangular cassette tapes, and square computer screens having replaced writing pads with the once familiar golden dimensions.

The survey experiment, though hardly definitive, challenged my assumptions about making reference to a "perfectly" proportioned rectangle and I had to let go of what was beginning to feel like aesthetic snobbery. Nevertheless, I returned to the architect's challenge of building a public structure, now with respect to the Parthenon. Master sculptor Phidias, I told my students, was commissioned by the Assembly of the Greek government to design a building that would signify Athenian ideals to the outside world, glorify the Acropolis, house the shrine of Athena, and of course, please the viewer. I emphasized that this was a publicly funded commission, so Phidias was obliged to design a public space with public appeal. Phidias, I continued, felt confounded by the task, and sought fresh perspective from his colleague, the mathematician Protagoras. (There is, to my knowledge, no specific account of how Phidias and Greek architects designed the Parthenon in the years 447–433 B.C.E., though we can assume that it was meant to accord with the understanding of human proportion generally credited to the sculptor Polyclitus. I thus took some liberty in narrating the Phidias story to my students, in order to give dimension to the process of discovering a design principle. I hope this won't ruffle the feathers of readers who may be better informed.) Protagoras happened to have just given a lecture on the Greek ideal and was delighted to offer his eloquent, functional statement, "Man is the measure of all things."

Taking the word *measure* to mean "basis," the statement can be interpreted independently from any link to architecture and the literal function of measuring. When I asked my students to interpret Protagoras' words, they said that Greek cosmology involved humanlike gods, and that the Greeks had high self-regard as a people who truly believed their civilization was more human than that of the barbarian. I wanted to nudge them further, toward an especially Greek exercise of translating a cultural ideal into a functional aesthetic. In the case of Protagoras' famous axiom, the idea that all things (of the natural world) exist in relation to Man, is also manifest in the discovery that the physical proportions of the well-built man, an obsession of classical sculptors, are those by which we can literally measure all things physical. I projected a slide of a classical *kouros* boy my students were familiar with from the Metropolitan Museum of Art, alongside Protagoras' famous words on the blackboard, and asked them to discuss the meaning of the quote in terms of architecture. This proving an insufficient prompt, I asked them to consider the forward-looking kouros boy in relation to the façade of a building. This finally yielded speculation: "The kouros boy looks forward like an important building." "There's symmetry in the body that an architect might want to imitate."

We were getting closer to objectifying the nude enough to see it in a purely structural way. Chalk in hand, I asked, "What happens if I draw rectangles around sections of the body like so?" I figured that they would notice the repeating ratio of the measurements of these adjoining rectangles if I made it sufficiently visual, but if I stimulated anything, it was deeper perplexity and a little frustration. I asked a student to stand before the class, legs and arms wide apart in Vitruvian fashion, and I proceeded to indicate rectangles around his hand, his forearm, his chest, then his torso. (Virtually all contiguous limbs and skeletal sections display the golden ratio.) Perplexity had now turned to hilarity and I had succeeded in making my model and myself feel rather foolish. Finally, I asked, "What about the ratio of this limb to that limb? Do you see that ratio in other regions of the body?" Their response was to ask, "What's ratio?"

Up to this point, despite the missteps any teacher makes when teaching new concepts for the first time, I had felt I was treading on firm ground within my discipline. Now I was in a situation in which it was necessary to cross the border, even if only temporarily. I felt a fresh respect for the art of designing lessons that parse terms and ideas that to the subject-area teacher seem self-evident. Could I relate the concept of ratio (which to me seemed fundamentally visual, a form of natural intelligence like two plus two) by continuing to focus on the human models we had before us, or did I need to clarify it in conceptual isolation? I rephrased my question: "Is this forearm to this hand, as this calf is to this foot?" Silence. Abandoning the question, which was, after all, a leading one, and perhaps unclear, I drew a couple of graphic puzzles involving proportions of shape (square $x$ is to square $y$ as circle $a$ is to circle $b$). They seemed to find these ratio puzzles pleasurable, and it occurred to me to mention that reasoning by analogy is a particularly Greek intellectualism wherein a conclusion emerges from intuition or "natural intelligence."

Calling my model back to the front of the classroom, I passed out fresh facsimiles of da Vinci's Vitruvius Man for my students to mark up and said, "Let's find dimensional ratios among the parts of this human form." As my students simplified the sections of the body into rectangles (Figure 9–1) they began to see a consistent dimensional ratio among contiguous sections of the body.

Now, my question was, "Could Phidias use this proportion to design his building?"; and, to return to the idea of a most preferred rectangle, one whose dimensions might well suit a wall or window, I asked them to take another look at the contiguous rectangles on Vitruvius Man and to isolate the total rectangles that most pleased them. We had returned to the matter of taste, only now the task of selecting seemed more reasonable, and when I indicated the rectangles deemed "golden," my students were more forbearing. I projected the original slide of the Acropolis and asked them to reexamine the Parthenon with the golden ratio and rectangle in mind. I replaced the slide with a simple diagram of the Parthenon's façade and asked a student to mimic the rectangle drawing he had drawn on the human body. He thought for a second, then drew two rectangles, one around the entablature and the other around the columns (see Figure 9–2):

An exciting moment—my students could now see the guiding principle behind this famous edifice, how the section-to-section ratio was clearly

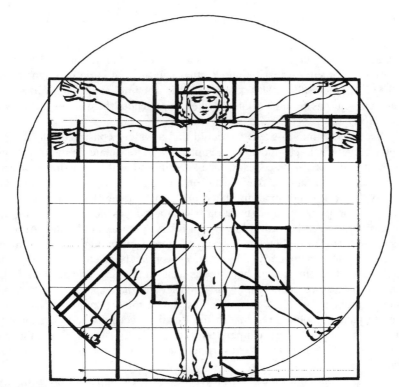

Figure 9–1.

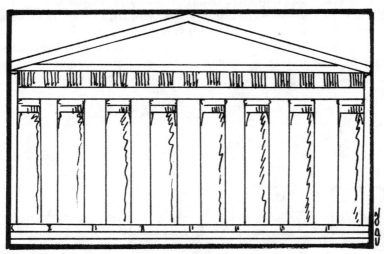

Figure 9–2.

derived from the human body. This was a synthesis of logic and aestheticism, a "Greek" moment, in that the intellectual pleasures of logic had transferred to the senses themselves. They knew viscerally the implications of Protagoras' statement—just as the ancient Greeks likely experienced them among their civic structures.

It was now time to see if we could find a way to construct golden rectangles ourselves. Each student contemplated a basic square on the board and on sheets of paper at their desks. One submitted that it was a simple matter of measuring the ratio of one set of contiguous sections from the Vitruvias Man with a ruler and applying that value to the square by either bisecting it at a certain point or supplementing it with a rectangle consistent with that proportion. I asked him to follow up on his suggestion and he did. Another suggested that we simply transfer the proportion by eyeballing it, and she managed to do so convincingly, adding an appropriately sized rectangle to the square. I spotted one student holding his hand to his face, fingers spread like calipers to measure the height of his forehead and the distance from his eyes to the bottom of his nose, then carefully (and I think jokingly) transferring hand to paper. I pointed out that in the Renaissance, followers of Leonardo da Vinci did indeed create calipers set to this ratio. Finally, I offered the traditional compass method for converting a square to a golden rectangle (see Figure 9–3):

I passed out compasses and they followed each step with no difficulty. Before, we had found ratios among different sets of rectangles around sections of the body. Now I could ask them to find a ratio strictly from this new subdivided rectangle. This challenge puzzled them, so I drew the subdivided rectangle on the board again and asked, "This smaller section is to this larger as *what* is to the total rectangle?" They were clearly perplexed by the absence of a fourth variable. This, I told them climactically, was the beauty of it: reapply the *larger section* to the whole, and you see the answer.

There's something to be said for allowing students to experiment and feel a trifle frustrated before intervening to provide an answer. Having said this, I

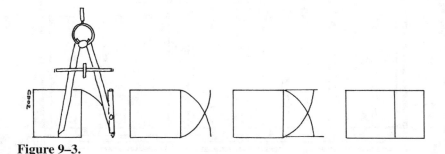

**Figure 9–3.**

suspect a geometry teacher would have found a methodical way for her students to deduce the sequence of steps for constructing the golden rectangle from a previously taught set of principles and to perhaps arrive at the resulting ratio on their own. Indeed, I passed what we had learned on to their math teacher who told me she would use it as a fine starting point for learning the accompanying math formulas. Nevertheless, it was enough for my students and me to marvel at the practicality of the compass method and the simple beauty of the self-referential golden ratio, having come to understand the rectangle's true derivation in the Greek ideal.

The second unit, "Universal Religions in the Middle Ages," was devoted to medieval Christianity and Islam. For our exploration of Christianity we would study replicas of two historical structures: the cloistered monastery and the cathedral. For Islam, we would visit a collection of tapestries, ceramic art, and a mosque. It was in the Islam component that I grappled with tessellation, a geometric form diametrically opposed to that of Classical Greece: based not on man, but on God.

I asked my students to characterize Islamic art based on slides of *mihrabs* (prayer niches), walls, and mosques. Islamic art, a blend of tessellation, arabesque, and Koranic verse, is ubiquitous throughout the Arab world, decorating every conceivable surface. The patterns themselves seem to spread beneath the boundaries of the wall or tapestry's edge to an infinite horizon. The main points I wanted to tease from their observation of these slides were that Islamic art is decorative, ubiquitous, and infinitely repetitive.

To make our shift into the world of Islamic aestheticism more clear, I projected a contrasting slide of a medieval French tapestry, *The Start of the Hunt*, the first of seven in the famous Unicorn Tapestries. I wanted them to articulate their awareness of the difference between narrative and abstract composition, the former being much easier to describe in words than the latter. Returning our focus to the Islamic forms, I gave them the terms *tessellation* and *arabesque*, and I pointed out a few of the architectural elements they overlooked, such as the hexagonal shape of a mosque made evident from a bird's-eye view, and the dome topped by a crescent moon. Finally, creating a bit of mystery to be explained later, I told them that the Arabs invented the numerical figures that we use to this day. I wrote a few numbers on the board and asked them to try to disassociate them from math class to see if they could simply see them as drawings: straight line, angle, and curve.

I asked for consensus that decorative surfaces we had seen were "pleasing to the eye." I think this final step, as self-evident as it may seem, is necessary for a lesson like this just as it was for the golden ratio lesson. Although we take for granted that patterns and symmetry are pleasing conventions, we should remember that someone at some point in history *discovered* such principles and *invented* ways of expressing them. Such discovery and invention are historical events worth ruminating on.

The next step was to return to the quest to identify a guiding principle. If in classical Athens the guiding principle behind art and architecture was "Man is the measure of all things," what was/is it for Islam? At first the question yielded very little. How could there be a guiding principle here? What could possibly be significant about these patterns? Curiously, more than one student suggested that they could spot fleeting human forms within the geometric patterns, prompting us to look closely for anything resembling the figure. Although the arabesque elements clearly rendered plants and flora, we simply could not identify human form among the other shapes. This led to a negatively phrased formulation: "Islamic art is art without human beings." An excellent start. We were headed for something clearly contrary to Protagoras' ideal. It just wasn't clear what we'd posit in place of "man."

I suspected that we could discuss the concepts behind Islamic art by creating our own tessellation. Students worked in pairs equipped with a pencil, a piece of string, and a blank sheet of paper. The paper represented a wall that they would have to cover with the sort of Islamic art they had seen. At first the task seemed absurd. A few groaned. "Where do we even start?"

Pondering this question, I remembered my first day of geometry in high school. (Geometry, perhaps not coincidentally, was the one area in the math curriculum that I actually enjoyed.) I remembered talking about the "cosmic point"—that one point was simply a point in space, but that two points, no matter where in the universe they lie, form a perfect line. One more point and you have a shape. Add another and you'll see three-dimensional form. Paul Zeitz, my teacher, took great pleasure in demonstrating this mystical movement from nothing to something. Here was an opportunity to introduce my students to the same discovery in the context of Islamic art. I suggested, in answer to their question, that they start with the simplest possible mark in the most obvious location: a dot in the center.

Students placed a second dot an inch or so from the first, measured the distance with a compass, and created a circle around the original dot that bisected the second. Our unit on Greece had touched on the virtues of angular geometry, and they learned about the functional arch from their textbook's discussion of Rome. Now we would consider the cultural/religious value of the nonfunctional circle. The question, What makes a circle meaningful? stimulated a thoughtful discussion of abstraction and perfection, concepts that the students would need to consider to delve more deeply into the values of Islam.

I asked my students to create an overlapping circle bisecting the center of the first. Once they had done so I asked them to identify the crescent moon shapes formed by blocking out the center oval. Now the task was to expand the design to five circles—that is, four circles, each bisecting the center of the original:

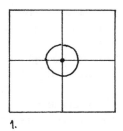 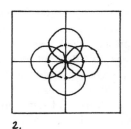 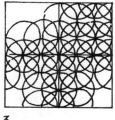

1.                    2.                    3.

Few could do this intuitively. Indeed, confusion set in when I asked them to bisect the original circle. It had become difficult to identify. The confusion was useful pedagogically, however, as it helped us grapple with our guiding principles. These five circles create a variety of patterns. I handed out color pencils and asked them to find, then shade in, the four-petaled flower in the middle of the design. They then bisected the original center point with two straight crossing lines. Referring to our earlier exercise on numbers, I asked them to identify as many as possible within this new composition of curved and straight lines. This process of discovering recognizable elements of nature and number seemed mystically gratifying.

To approach our guiding principle, I asked, "What is the significance of starting with a dot in the middle of the page (or wall) and continuing outward? What does it mean that after you repeat identical, overlapping circles several times, it becomes nearly impossible to identify the point at which you started? What does it mean that the shapes want to continue outward beyond the edges of the page, that indeed there's no way to cover the space without cutting shapes off along the borders? What does it mean that we were able to create patterns that resemble elements of *nature* by repeating these shapes?"

The synthesis of all of these ideas is grandiose and perhaps a bit unfair to ask of the average fourteen-year-old. I remembered reading Descartes and Spinoza in college and getting to those points in the text where the existence of God is posited through rational induction. Here we were headed toward an elusive analogy between art and Allah. Could my students see that their tessellation was imbued with mystical symbolism?

On the blackboard I wrote a famous syllogistic line from the Koran: "God is beautiful and loves beauty." I then erased *is*, and replaced the word with an equal sign, and for good measure, asked my students: "Are these geometric shapes beautiful?" then, Socratically, "Insofar as the shapes repeat themselves, and spread themselves over and over, can it be said that they love themselves?" Although they didn't use the exact phrasing of the understanding I wanted them to achieve, that "Islamic art enacts an infinite self-beautifying, self-reproducing

Allah," the class did come away with a profound *sense* of the connection between Islamic art and Islamic theology. Certainly, we had delved into an abstract realm in which imagery is not representative but synonymous with an idea.

Both lessons in our two curricular units engaged elements of design that to the modern viewer may seem plain and familiar: symmetry, proportion- ality, and pattern. Western civilization sustained Grecian architectural design through neoclassicism and appropriated Islamic art for all sorts of things, from neckties to wallpaper. The question is, Can we disaccustom ourselves from such predictable and familiar aestheticism to apprehend its enormously meaningful origins? As an *ad hoc* geometry teacher in a humanities classroom, I believe I can safely say, "Yes," for I discovered that although those simple lines, squares, triangles, and circles combine to content the eye, their essential nature, once unlocked, invigorates the mind.

# Work Cited

Fechner, G. 1860. "Rectangle Preference Survey." From handout by Michael Schneider, *Greek Art and Mathematics* (Metropolitan Museum of Art: Education Department, 1993.)

# Bridge

I met Avi in 1998, my first year as a staff developer in District Two. He was teaching a literacy class of sixth- and seventh-graders in summer school, using the balanced literacy model. During a minilesson on punctuation he instructed the students to put their fingers on the relevant spots of a shared text on their desks. The sense of combined focus·was palpable, as if the usual distracted energy associated with this population had been put through a lens to create even greater heat than in a normal classroom. Grammatical usage rapidly improved during work time.

As an English and humanities teacher in the school year, Avi becomes an authority on what he learns from his far-ranging research. He creates engaging projects for his students but he can also lecture if the occasion calls for it. I once had the pleasure of watching him spellbind a heterogeneous group of students, some of whom had notable attention problems. In a courtyard of the Metropolitan Museum of Art he addressed them on the history of the arch, where they would find various examples in the museum and where they could find further examples in familiar neighborhoods in the city. After his talk the students dispersed to work on their architectural inquiry projects like children on an Easter egg hunt.

Rereading Avi's essay I get the pleasant sensation (speaking of sensations) that even as the plot of our book turns back toward its point of origin, it is spiraling ever deeper into our theme. Math and the humanities are married in his examination of the aesthetics of design. (Few but Avi would have the imagination to speculate that the shape of CD jackets might be the reason teens don't respond to the golden rectangle.) They are married in his extraordinary response to the need to teach ratio and in the depth of his passion to teach what Greek logic feels like, in one's intuition, in one's "natural intelligence." Aristotle—an early constructivist?—articulated this in his poetics when he said, "There is nothing in the intellect that is not first in the senses."

Implied by the essay is a comparison between Greek culture B.C.E. and the medieval culture of the Muslims. The Greek fear of irrational measurements was matched by their fear of the infinite (the Latin term for which is *horror infiniti*—which makes me imagine them wearing garlic and holding crosses up to the square root of two). The Muslims, by contrast, found the infinite liberating in its irrationality. The way Avi led his students to this understanding of a fundamentally mathematical issue through a series of humanistic questions wonderfully illuminates our theme: What is the guiding principle in Islamic Art? What comes *in place of* man? In constructivist fashion, the students had to investigate the questions for themselves, through observation and hypothesis.

They will never view circles the same again when they go back to math class, not after, as Avi put it, having such a "mystically gratifying" discovery of its aesthetic and religious function for Islam.

Avi's essay actually reveals the humanistic bias of our theme. To math students, it may well be the *humanities* that feel like "the other." It must have made such students feel less alienated then to use the instruments of mathematics in a humanities class. Indeed, without the math, the human dimensions of this study would have been woefully incomplete. But how many humanities teachers would have had the confidence to do what Avi did? Of course, it is our hope that this book will increase that number.

Another teacher who experimented successfully with math in her humanities class is Amy Samson, who taught sixth grade at The Clinton School for Writers and Artists when she wrote her chapter. She happened to also be working with the topic of the Middle Ages.

# 10

## Scaling in Humanities

*Expanding Confidence, Engagement, and Understanding*

### Amy Samson

The Clinton School for Writers and Artists, a middle school in the Chelsea neighborhood of New York City's District Two, has a distinguished history of successfully integrating social studies, visual art, and language arts in a humanities block for its sixth graders. I loved the convergence of subjects. It made sense both to my students and to me to combine those areas of inquiry. I routinely greeted my students as they came from their math classroom across the hall while sending my other group of students to that same math classroom. I knew that the math teacher, Christine Dorosh, was having her students keep math journals. I began to wonder how to bring mathematics across the hall and integrate it into my humanities block. Was there a sensible way for an English teacher to create a bridge between *math* and humanities?

I explored ideas for accomplishing this in discussions with Christine. My students already used some mathematical concepts to plot a timeline of the European experience from Ancient Greece to the Middle Ages. Christine and I had also planned and implemented a short unit that combined the study of stained glass in the Middle Ages with the design and production of miniature stained glass windows for our classroom. Small groups of students chose themes or stories and used tessellations, a concept they learned in their math class, as the underlying design element to unify a series of windows. These projects were rewarding, but I wanted to integrate mathematics more completely into a unit of study.

In previous units I had discovered that my students' understanding of scale and distance was an area of weakness. This had been a barrier to fully comprehending the geography of Egypt and Greece. I was about to begin my unit on the Middle Ages, which would again raise issues of scale and distance; however, my students wouldn't encounter scale in their math curriculum until the seventh grade. Could I take this on myself? With encouragement from Christine, I decided to try, and embarked on what became a deeply fulfilling literary, historical, cultural, and mathematical journey through an important historical period.

I planned to start this unit, as I did all my units, by working from the students' own questions about the subject at hand. Often the questions they generated coincided with material I'd planned to include already. Just as often, they raised questions I had not considered. I began by discovering what they already knew and finding out what they wanted to know. What do you know about Europe in the Middle Ages? and What do you want to know about Europe in the Middle Ages? were the questions I posed.

Their answers displayed broad knowledge and curiosity about the Middle Ages. Some students wrote about France or England, some wrote down dates for the Middle Ages, and some revealed knowledge about technological and artistic advances made during the Middle Ages. Other students questioned exactly what the Middle Ages were and when and where they had occurred. Our discussions led us to focus on Europe between 1000 and 1485 C.E.

To begin an exploration of the geography of the continent, students worked in groups with many types of atlases. I provided templates for them to create maps with boundaries that existed in 1300 C.E. Figure 10–1 shows my project description.

Figure 10–2 is the map template that I used:

Students added geographical features, country and empire boundaries, a compass rose, and longitude and latitude lines. Once students had completed their maps, we were ready to begin our discussion of scale.

I explained that we would read a well-researched historical novel set in 1299: *The Ramsay Scallop*, by Frances Temple. The characters in the novel travel on a journey throughout Europe, beginning in a small village in England and ending in Spain. As a class, we would create a wall-sized map to trace their journey. While they were excited by the prospect of a large-scale art project, the students had questions about how we would accomplish it. When I told them that we would use their math skills to create the map, their enthusiasm wilted. Some of my strong writers and readers showed a sudden lack of confidence. I had anticipated a certain degree of resistance to doing math in humanities, but the intensity of their reaction still surprised me. It thus became one of my unit goals to have them gain the confidence I associated with meeting the NCTM (National Council of Teachers of Mathematics) standards of performance. To reassure them, I pointed out that all of them had successfully created maps of Europe in the year 1300 C.E. We simply needed to enlarge those maps to create our classroom map for display. We would figure out the mathematical process together.

Humanities 601/602

## MAPPING MEDIEVAL EUROPE

The attached map contains parts of Europe, Asia and Africa. Using the books on your table and colored pencils, complete the following steps to create a map of Medieval Europe around the years 1000–1300 C.E.

1. *Title your map "Medieval Europe, 1000–1300 C.E."* Write the title in the box in the upper right corner of your map.

2. *Make a key for your map.* Use blue for water and brown for mountains. Create a symbol for cities.

3. *Label the following in legible handwriting:*

| | | | |
|---|---|---|---|
| Bodies of water: | North Sea | Mediterranean Sea | Atlantic Ocean |
| | Black Sea | Adriatic Sea | English Channel |
| | Ionian Sea | Aegean Sea | Bay of Biscay |
| | Baltic Sea | Strait of Gibraltar | |
| Draw and label the following rivers: | Rhine River | Danube River | Seine River |
| | Loire River | Elbe River | Oder River |
| Mountains: | Pyrenees | Alps | Carpathian Alps |
| | Dinaric Alps | Apennines | Balkan Mountains |
| Continents: | Africa | Asia | Europe |
| Regions: | France | Spain | Italy |
| | Germany | England | Greece |

Cities: London, England
Paris, France
Bordeaux, France
Rome, Italy
Santiago de Compostela, Spain
Constantinople (present-day Istanbul, Turkey)

4. *Add a compass rose to your map.*

5. *Create a distance scale for your map.* Place your distance scale in the appropriate place.

**Figure 10–1.**

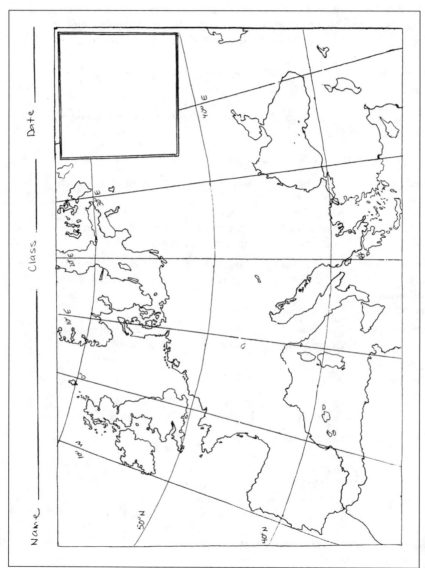

**Figure 10–2.**

To model the process, I drew a chevron-shaped hexagon on an 18-inch square of white paper:

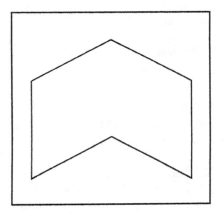

I asked for a student volunteer to reproduce the figure on a 36-inch square piece of paper. I asked the class what next steps might be effective for creating an accurate enlargement. By guiding and questioning one another, we decided to create matching grids on each square. We folded each square in half and then into quarters. Then, we used a ruler to trace the fold lines in ink. The students' excitement was palpable when they realized how simple it would be to create an accurate, proportional enlargement of the hexagon. Several volunteers stepped forward to create the enlargement. They drew one square at a time, using the grid lines to determine where to place lines of the figure that fell within their square. In a short time, they had a 36-inch reproduction of the original hexagon:

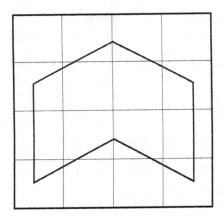

The ease with which they completed the process gave all the students the confidence they needed to proceed.

Before moving on to their maps, which they all wanted to do immediately, we reviewed the process of creating our enlarged polygon. On chart paper, we created a list of steps for enlarging figures.

1.  Fold your original to create a grid of sixteen boxes.
2.  Fold your large, blank square to create a grid with the same number of boxes.
3.  Carefully enlarge each square, using a *pencil*. Use grid lines for reference.
4.  Write the word *top* in pencil at the top of your square.
5.  Write your name on the back of your square.

I selected several complete, legible, and accurate maps from among my students' original small-scale maps. Using a photocopier, I enlarged them to 11 by 17 inches. I drew a grid of 2-inch squares over the clearest map and numbered each square to guide the assembly of the map. Once I drew the grid and numbered the squares, I made a few copies of the map for reference. Last, I cut the map into squares and put them in an envelope for distribution to my students.

Before beginning the process of enlarging the map squares, we reviewed the steps on the chart paper. I distributed a 2-inch square from the map and a blank 12-inch square to each student. I asked students how large the finished map would be once the enlarged pieces were all assembled. Since they knew the sizes of their small squares and the blank squares, I asked them to talk to the students around them in their groups to either predict the size of the large map or generate a list of questions to figure it out. After some time, a student asked me the size of the original map before I cut it into squares. When I told them that the original map was 10 by 16 inches, they predicted that the large map would be 60 by 96 inches. When students realized that the large map would be 8 feet long, they could hardly contain their excitement.

At this point, I reminded students to follow the steps on the chart paper and begin working independently to enlarge their sections. I stressed the importance of accuracy and care, since we would have to fit all the pieces together like a puzzle. They worked with a startling level of concentration. I was able to monitor their work and provide guidance to students when necessary. I provided additional squares for students who finished their first squares while their classmates were still working. We had begun reading *The Ramsay Scallop* by this time, so once all forty squares had been distributed, students read silently as they waited for their classmates to catch up.

In the next session, to allow for the flexibility necessary to complete the map, students gathered in their groups and visited workstations around the perimeter of the room. The stations consisted of six tables with sources of information about the different aspects of the Middle Ages that had arisen in their original inquiry: knights, education, daily life, religion, art and architecture, and the crusades. Tasks at each station included note taking, drawing, and guided responses to the content. Students would use the information they learned at the stations to enrich their experience with the novel. This information would also enable them to create more realistic characters for the medieval fictional biographies they would write following the novel study. The final station was a large area on the floor in the middle of the room where the map would be assembled and painted.

I instructed the first group of students at the map station to order, assemble, and join the squares for the large map. Students used the numbers on each square and the 11-by-17-inch master map to accomplish this and to check the accuracy of the map. They verified the spellings of landforms, countries, and empires; they checked the lines of longitude and latitude; they redrew any lines that did not quite meet at the places where the squares were joined. Once the map was assembled and accurately drawn, the students at this station began to paint. They created a key and decided what colors they would use. They would paint England gold, France purple, the Holy Roman Empire dark green, Burgundy yellow, and Spain light green.

Groups moved to new stations every thirty minutes. The group that had been working on the map brought the next group up to speed, giving me the freedom to provide assistance to students at the other stations. This enhanced students' communications skills, since they were entirely responsible for explaining the status of their work and guiding the next group of students. Once the map was complete, we hung it on a wall and used it to chart the progress of Elenor and Thomas, the main characters from *The Ramsay Scallop*, as they journeyed to Santiago de Compostela. When the young travelers came to new towns, we used an atlas to find the coordinates of longitude and latitude needed to place the towns on our map. When they crossed rivers or climbed mountains, we had a constant visual reminder of their progress and the difficulties they encountered.

We had reached our goal of using scale to create a wall-size map of Europe. But I wasn't certain my students had a true understanding of the distance the characters in the novel traveled. To relate the characters' journey to a distance that my students knew, we took a familiar distance and estimated how long it would take to walk that distance. The distance from the Clinton School, on West 21st Street, to the Museum of Natural History, on West 79th Street, is approximately $3\frac{3}{4}$ miles. We had traveled that distance by subway on a field trip earlier in the year. Based on the amount of time it took students to walk several blocks, we estimated that it would take us nearly an hour to walk the distance to the museum. Next, we calculated the distance from Paris to Bordeaux using the distance scale on their map. Respect for the courage of young people in the Middle Ages was visible in my students' faces when they realized the effort that journey entailed.

My goal of creating a better understanding of scale and distance turned out to be a relatively minor accomplishment in this unit. Of far greater importance was the level of engagement that led to that understanding. I could now send my students back across the hall to their math class secure in the knowledge that math belonged in my humanities class as much as the writing associated with the humanities belonged in math.

# Resource Books

These are adolescent books and atlases that I used for the stations.

Adams, B. 1981. *Atlas of the World in the Middle Ages*. New York: Warwick.

Ancona, G. 1995. *Cutters, Carvers & the Cathedral*. New York: Lothrop, Lee & Shepard.

Bard, I. 1989. *History of the World: The Late Middle Ages*. Milwaukee: Raintree.

Batterberry, M. 1961. *Art of the Middle Ages*. New York: McGraw-Hill.

Black, I. S. 1963. *Castle, Abbey, and Town: How People Lived in the Middle Ages*. New York: Holiday House.

Briquebec, J. 1990. *The Middle Ages: Barbarian Invasions, Empires Around the World and Medieval Europe*. New York: Warwick.

Brooks, P., and N. Walworth. 1966. *The World of Walls: The Middle Ages in Western Europe*. Philadelphia: Lippincott.

Caselli, G. 1988. *The Middle Ages*. New York: Peter Bedrick.

Clare, J. D. 1993. *Fourteenth-Century Towns*. San Diego: Gulliver.

Corbin, C. 1989. *Knights*. New York: Franklin Watts.

Cosman, M. P. 1981. *Medieval Holidays and Festivals*. New York: Scribner's.

Dawson, I. 1994. *Food and Feasts in the Middle Ages*. New York: New Discovery.

Gibb, C. 1985. *Richard the Lionheart and the Crusades*. New York: Bookwright.

Glubok, S. 1969. *Knights in Armor*. New York: Harper & Row.

Gravett, C. 1994. *Castle*. New York: Knopf.

Gravett, C. 1993. *Knight*. New York: Knopf.

Knight, J. 2001. *Middle Ages Almanac*. Detriot: UXL.

Leaon, V. 1998. *Outrageous Women of the Middle Ages*. New York: Wiley.

Macaulay, D. 1973. *Cathedral: The Story of Its Construction*. Boston: Houghton Mifflin.

———. 1977. *Castle*. Boston: Houghton Mifflin.

Macdonald, F. 1984. *Everyday Life: The Middle Ages*. Morristown, NJ: Silver Burdett.

———. 2000. *Women in Medieval Times*. Lincolnwood, IL: Peter Bedrick.

Macdonald, F. and M. Bergin. 1990. *A Medieval Castle*. New York: Peter Bedrick.

Macdonald, F. and J. James. 1991. *A Medieval Cathedral*. New York: Peter Bedrick.

Nikola-Lisa, W. 1997. *Till Year's Good End: A Calendar of Medieval Labors*. New York: Atheneum.

Riley-Smith, J., ed. 1990. *The Atlas of the Crusades*. New York: Facts on File.

Sancha, S. 1982. *The Luttrell Village: Country Life in the Middle Ages.* New York: Thomas Y. Crowell.

———. 1987. *Walter Dragun's Town: Crafts and Trade in the Middle Ages.* New York: Thomas Y. Crowell.

Sellman, R. R. 1955. *The Crusades.* New York: Roy.

Smith, B. 1988. *Castles.* New York: Franklin Watts.

Sobol, D. J. 1959. *The First Book of Medieval Man.* New York: Franklin Watts.

Suskind, R. 1967. *Cross and Crescent: The Story of the Crusades.* New York: W. W. Norton & Co. Inc.

Tappan, E. M. 1911. *When Knights Were Bold.* Boston: Houghton Mifflin.

# Bridge

A picture on my desk shows Amy hiking in the mountains near her home in Eugene, Oregon, where she moved after the year I worked with her at the Clinton School. Among the mountains she is a tiny distant figure in a red shirt, wearing a weighty blue backpack and dark glasses. Small as she is in the photograph, she seems to be looking directly at me, across a gorge the size of the United States, as if to say, "What are you doing back there at your desk in the city, you fool? There are other ways to spend your life, you know." Alas, until this book is finished, I will have to participate in those other ways vicariously. Luckily, Amy has done a wonderful job of making that possible. Here is a particularly poetic excerpt from our correspondence:

> Wes and I went snowshoeing in the Cascades this weekend. I rowed early in the morning (another magical morning—glassy water, a juvenile bald eagle who kept flying over us, a fine mist that cooled us down and gave rise to a rainbow), and Wes picked me up and we drove just thirty minutes to get from green grass to six feet of snow on either side of the road. We chose a lightly traveled trail up Fuji Mountain and hiked until we couldn't go any higher. We stopped to eat lunch, enjoy the sun, and built a smiling snowman to leave behind and descended with a view of the surrounding mountains all the way down. Do you think you could get funding to bring the special ed kids out here? We could do a hiking expedition/writing workshop with them!!! :)

That's Amy. A person to listen to when she talks about teaching, or anything else for that matter. She should certainly be listened to when she talks about teaching. Digesting the thoughts in her essay, we see how closely it cleaves to our theme. It underscores the assumption we too frequently encounter among teachers that it's easier to incorporate visual and language arts into humanities than math; and students share this assumption, as we see from their dismayed reaction to the idea of using math to make their map. Maybe it's the word itself: *math*. Sounds like the bleat of a goat. And the association with *after*math is not a happy one. Certainly, when the task itself was broken down the students couldn't have been happier than doing this bleaty math. The essay shows how the assumption is false, how on closer examination the line between math and the humanities is very blurry; but it also demonstrates how, as with Avi, taking some responsibility for knowing what your students need mathematically as well as historically and linguistically yields benefits in all three areas.

The designated math skill that Amy was willing to teach, scaling, was not the only math benefit that the students derived from the lesson. Amy knew her math standards. She knew that these, in particular, were addressed by her project:

reasoning proportionally, estimating and comparing magnitudes, making predictions, estimating numerically, measuring accurately, referring to geometric shapes correctly, using mathematical language, organizing work, explaining solutions orally and in writing, organizing data on charts and graphs, explaining ideas to others, studying data, modeling mathematically, designing physical structures, management, and planning. She knew the natural correlation with ELA standards, and she knew they would flow naturally from the project design.

Amy also knew her constructivist methodology. She elicited what the students already knew, had them articulate what they wanted to know, and assessed what they eventually learned. She was truly a facilitator as the students "guided and questioned one another." "I was able to monitor their work and provide guidance to students when necessary," she said. The geography of her room reflected her style as much as the geographical content of the project, with the central artwork occupying the center of the room and the research stations positioned around the edges, making the rotation from group to group flow easily.

In the end what she valued most, though, after all the math skills, was that the students were "secure in the knowledge that math belonged in the humanities class as much as the writing associated with the humanities belonged in math," and she proved it not only with the map project but also with the tessellation project that so eerily complemented Avi's work. I found it fascinating how she synthesized the narrative art of Christian Europe with the nonrepresentational Islamic art of the Middle Ages, by using the tessellations as the underlying design linking themes and stories.

Elizabeth Fox examines the nature of narrative and nonnarrative forms in our next chapter.

# 11

## Encouraging Chaos

### Elizabeth Fox

I sit by the window in a coffee chop on Chambers Street in New York City. As people walk by some look in the window and others look straight ahead. Chinese student, straight ahead; green parka reads a newspaper; beret sidelong glance inside; red shawl straight ahead; fashionista in a fuzzy cap looks in and up at the ceiling; working mother hugging a packet of folders, straight ahead; blue ski cap and matching backpack strolls by looking in with curiosity; lawyer in black trench coat, straight ahead; African American with kenta cloth crown, straight ahead; gray beard quick glance in then straight ahead; straight ahead; straight ahead; then students, cashmere and baseball caps, slow stroll looking in; youthful professional, straight ahead.

I love the variety of people walking by on a winter's day. Who can predict who will next appear? Why they chose the coat, scarf, and hat they wear? How they came to be making deals on a cell phone, remembering lines from Chaucer, or imagining the perfect apartment? Directing my attention toward the people and quirks of my surroundings is part of my writing routine. I mine ordinary scenes for anything that might quicken my delight for the chaotic way life is conducted.

My appreciation as a writer of the random events around me has a corollary in science. In the study of chaos, scientists and mathematicians have noticed that order arises spontaneously in complex, irregular systems. In *Chaos* James Gleick writes,

> Where chaos begins, classical science stops. For as long as the world has had physicists inquiring into the laws of nature, it has suffered a special ignorance about disorder in the atmosphere, in the turbulent seas, in the fluctuations of wildlife populations, in the oscillations of the heart and the brain. (1987, 3)

Classical, Newtonian science is based on a deterministic model of the universe. At its foundation is the belief that if it were possible to know the exact location, speed, and weight of every subatomic particle in the universe, then it would be possible to predict what would happen next. A cause leads inevitably to a predictable effect. The quantum physicist Werner Heisenberg demonstrated early in the twentieth century that this understanding didn't apply at the subatomic level of physical reality. He showed that it is not possible to observe both the velocity and the location of a subatomic particle at the same time. Determinism was thus an incomplete description of how nature works. Toward the end of the century, the study of chaos emerged to fill the gap in understanding created by Heisenberg. Instead of predicting exactly what will happen next in a system, the science of chaos seeks to understand a creative dimension of nature. Gleick explains,

> Those studying chaotic dynamics discovered that the disorderly behavior of simple systems acted as a *creative* process. It generated complexity: richly organized patterns, sometimes stable and sometimes unstable, sometimes finite and sometimes infinite, but always with the fascination of living things. (1987, 43)

Writing teachers often use a deterministic model when teaching writing. They may, for instance, teach students to use the prescribed form of a five-paragraph essay to complete an assignment: first paragraph—introduction and thesis statement; second—first supporting statement and evidence; third—second supporting statement and evidence; fourth—third supporting statement and evidence; fifth—restatement of thesis and conclusion. Many English teachers also use a deterministic model when they teach high school students how to read literature. They encourage their students to discern the author's thesis and to explore how the author's formal choices support the thesis. This analysis may result in the development of a belief that writers always write in a deterministic way by working out their thesis, then writing with the intention of making their characters symbolic and investing every detail with layers of meaning.

Just as determinism works up to a point in physics, so it does in reading and writing. If I push my cup of coffee off the edge of the table, I can predict that it will fall. In writing, formulaic writing is quite serviceable for thousands of professional writers. But classical physics is insufficient to predict the effects of my spilled coffee, and a deterministic model of teaching writing doesn't admit the possibility of disorderly behavior in the creative process. It may yield work that is competent but lacks life.

I find this to be the case when I work with students in the Writing Division of Columbia University's Summer Program for High School Students and with students attending Columbia Scholastic Press Association conferences. I want them to experience the secret delight of leaving the predictable, formulaic world they have known to discover a new world of possibility in language. I want them to have an experience that is linked to a sense of process rather than state. Instead of thinking of themselves as writers who impose their will and realize their conscious intentions on their work toward a specific result, I want them to be explorers in language.

Because I usually see my students only briefly, for as little as one forty-five minute period to no more than sixteen sessions, I try to introduce them to techniques that exaggerate the stochastic processes. Many stem from the traditions of Dada and surrealism that seek to bypass conscious self-monitoring and internal judges. Others reveal the potential for literature that exists when the writer's conscious intention is intentionally thwarted.

One excellent source of writing ideas that circumvent deterministic writing practices is the work of the Oulipo, a literary group based in France whose members include writers and mathematicians. Oulipo is derived from its first name, Ouvroir de Littérature Potentielle (Workshop for Potential Literature). According to one member, Jacques Roubaud, "The aim of the Oulipo is to invent (or reinvent) restrictions of a formal nature (constraints) and propose them to enthusiasts interested in composing literature" (Matthews and Brotchie, 1998, 38). A prime source of restrictions for the Oulipo is mathematics. For example, Paul Braffort wrote *Mes Hypertropes* (My Hypertropes) based on the sequence of Fibonacci numbers (each number is the sum of its two predecessors) and Zechendorf's theorem.

One assignment I've developed is derived from a constraint called "100,000,000,000,000 Poems" that was invented by one of Oulipo's founders, the writer Raymond Queneau. Queneau wrote a sequence of ten fourteen-line sonnets in which each line from any sonnet can replace the corresponding line in any of the other sonnets, resulting in the possibility of $10^{14}$ poems.

For one of my workshops I modified the constraint so that it could be done within a forty-five minute period by changing the form from a sonnet to a cinquain. According to the *Teachers and Writers Handbook of Poetic Forms*, "A cinquain (from French meaning 'a grouping of five') is either a five-line stanza or a poem with five lines. The cinquain has five lines, with two, four, six, eight and two syllables respectively (Padgett 1987, 48)."

I started the session with a brief explanation of Oulipo and announced that the six of us would produce 46,656 ($6^5$) poems in the next half hour. The four students and their teacher were skeptical about the process, and understandably doubtful that they could produce so many poems so quickly. After defining and showing them models of cinquains, I invited them to try their own. Then I had them transfer the individual lines of their poems onto colored index cards: the first lines of two syllables onto blue cards, the second lines of four syllables onto pink cards, and so on. After we listened to one another's poems, I collected and shuffled the cards to create new cinquains. When the writers heard their lines appear in new poems, they sensed the beauty inherent in a chaotic view of nature. The pleasure was evident on their faces.

To get an idea of our experience, read these three cinquains first as they were written, then pick a different line from each column to create a new cinquain. From these three cinquains it is possible to read 240 others.

| Ribs now | Moon beams | Love can |
|---|---|---|
| and the rise of | Light the night sky | Do wonderful |
| air, together combine | All sleeping, none stirring | Things to a person's life |
| like solid branches or wind chimes | As the earth turns and moves onto | Live each moment as if it was |
| inside | Tomorrow | Your last |
| *Jonathon Hilpert 6th grade English teacher Scarsdale Middle School Scarsdale, New York* | *Shoshana Schwartz Sophomore Cresskill High School Cresskill, New Jersey* | *Jennifer Sinisi Freshman Cresskill High School Cresskill, New Jersey* |

# Works Cited

Gleick, J. 1987. *Chaos*. New York: Viking.

Mathews, H., and A. Brotchie, eds. 1998. *Oulipo Compendium*. London: Atlas.

Padgett, R., ed. 1987. *The Teachers and Writers Handbook of Poetic Forms*. New York: Teachers & Writers Collaborative.

# Bridge

Many features of Elizabeth's exercise appeal to our theme. The overarching awareness of the math-based science and the particular application of math to literature enhance understanding of concepts in both math and the humanities. Students cannot help but become more sharply aware of the way chaos and order are embedded in the nature of reality through their experiments with its literary as well as its mathematical expressions. The exercise also nicely mirrors the work done with poetry in a *math* class by Matthew Szenher and me. In both cases, it's interesting to compare the feelings of the students at the outset with their feelings at the end. The doubts that math and poetry have much to do with each other almost invariably vanish.

How can it be, though, that the seemingly deterministic thesis of this book accommodates a practice relying on the randomness in chaos? It's a paradox, and conforming to the central point of our thesis, paradoxes have the potential to give pleasure, if we are able to open our minds far enough to contain them. The word has Greek roots that mean "contrary to received opinion" or "outside of dogma." "Form is emptiness, emptiness is form," say the Zen masters. "Do I contradict myself? Very well, I contradict myself," says Whitman. "Consistency is the hobgoblin of small minds," says Fitzgerald. How can we live, I ask, if we can't take pleasure in paradox?

I am tempted here to investigate the history of paradox in mathematics. I consult *The HarperCollins Dictionary of Mathematics* (Borowsky and Borwein 1991, 433) and find descriptions of famous paradoxes from logic and set theory. Some, I discover, such as Russell's paradox and Cantor's paradox, require "revision of the intuitive conception from which the paradox was derivable." Others "depend upon the strict inadmissability of the description of the paradox." Yet others "merely draw attention to some feature of the formal theory which is contrary to intuition." Discovering and resolving (or living with) paradoxes forces us, as Elizabeth puts it, to "admit the possibility of disorderly behavior in the creative process."

Admitting chaos is important not just for writing poems but for learning math, as well. It is the food of intuition. As Dehzene points out in The Number Sense, "The flame of mathematical intuition is only flickering in the child's mind; it needs to be fortified and sustained before it can illuminate all arithmetic activities. But our schools are often content with inculcating meaningless and mechanical arithmetic recipes into children (1997, 139)." So it goes with writing, as well, Elizabeth suggests. Whether humans' number sense is independent of linguistic function in the brain or not (it is in some respects, in others not, as we discussed previously), it is clear that the life of the intuition is necessary to the development of both. It is also clear that the intuition is as fragile as life itself, and equally in need of a nurturing environment.

Exactly how do we go about *providing* a nurturing environment? The question leads us to the last section of the book, which discusses the mathematics of engagement, and to the friendly seventh-grade classroom of Matt Wayne.

## Works Cited

Borowsky, E. J., and J. M. Borwein. 1991. *The HarperCollins Dictionary of Mathematics*. New York: HarperCollins.

Dehzene, S. 1997. *The Number Sense: How the Mind Creates Mathematics*. New York: Oxford University Press.

# 12

## Getting Smarter

### A Seventh-Grade Class Researches and Reflects on Its Discussion Habits

### Matt Wayne

> Intelligence: "the ability to learn or understand/the ability to
> problem solve"
> Intelligent: "having or showing intelligence"
> —Webster's New World Dictionary

Let me get this straight. Intelligence is the ability to learn and to solve problems. So individuals are intelligent when they have intelligence, right? But wait a minute. Doesn't everyone have the ability to learn and understand? That's the premise of the current new standards movement in education: "all children can learn (Institute for Learning 1999, ii)"; but much of our talk about intelligence both inside and outside schools doesn't reflect this idea. Most of the time, being smart is a "done deal." Melinda's smart; Catherine isn't. They're just made that way. It doesn't change.

Unfortunately, many children have already been labeled this way by the time they reach middle school. Some are tracked into "gifted" or "mainstreamed" classes. In one school, students are either in the "rainbow" class or the "real-life" class. How can children avoid wondering if they are the "intelligent" ones or not? Although kids most certainly have different abilities, doesn't every child have the capacity to learn? To become smarter? Yet if we treat intelligence as an innate quality, we limit their vast potential for learning and don't acknowledge or respect their efforts.

For the past twenty years cognitive and educational psychologists have been developing the idea that people become smarter. The Institute for Learning at the University of Pittsburgh calls the notion "socializing intelligence." Headed by Lauren Resnick, the Institute for Learning developed the "Principles of Learning" as a theoretical framework for the kinds of practice and environment one would find in a successful classroom. It defines socializing intelligence as "much more than a collection of bits of knowledge and quick reasoning tricks. Intelligence is equally a set of beliefs about oneself—one's right and obligation to understand and make sense of the world, and one's capacity to figure things out over time (4)." The principle goes on to term what it calls "functional intelligence" as "both a set of problem-solving and reasoning capabilities and the habits of mind that lead one to use them regularly (4)."

The foundation of the learning in my seventh-grade English language arts class is the belief that all children can develop intelligent habits of mind. In our class, "intelligent" is something you become, not something you already are. This is the story of how our class became a community to solve the problem of how to become better learners. In other words, this is the story of how my kids became smarter students and I became a smarter teacher. Along the way, it will become evident that mathematics, in the form of data, was essential to the process.

## The Power of Research

I began the school year eager to dive seriously into our reading and writing work. My class is an inclusion class at a small middle school in New York City's Community School District Two in which special education students participate in the general education classroom. Many of the students are considered "at risk." A majority of the students come from the nearby housing projects and qualify for school lunch. Located near Chinatown, the student population is roughly one-third Asian, one-third African American, and one-third Latino. I wanted to begin the year getting to know my students and creating a strong sense of community through readings and discussions. I chose to read aloud several stories on themes that I thought would interest them, such as "friendship," "school," and "growing up." I was hoping that the discussions after the read-alouds would offer me insight into our class community.

On the second day of class I read "Thank You, Ma'm" by Langston Hughes, a short story about a young boy who unsuccessfully tries to steal a woman's purse. The woman feeds the boy dinner, gives him the money he wants, and teaches him an important lesson. We began sharing our ideas after agreeing on the basics of what makes a good discussion: listening respectfully, taking turns, and supporting your ideas. Here is our first conversation after the reading (students' names have been changed):

> "What do you think about the story?" I ask an open-ended question to allow any ideas to flourish. Silence. I give them time to think. . . .
>
> "Why did Roger do that?" Kathleen asks about a character from the story.
>
> "That was wack."
>
> "I know, he's messed up," replies Tyrell.

"Oh, and can you believe that she did that to him," someone says, but I'm not sure who it is because a lot of students begin talking at once.

"Yea, it's like that episode of *The Fresh Prince of Bel Air* I saw last week . . . ."

"That show's stupid. I like *Sister Sister*."

"One at a time," I say, my voice carrying over the students. "There are some good ideas here, but let's try to listen to each other and stay on the topic. Remember, we're trying to have a conversation."

My interjection quiets the students down. Okay, I think, let's try to focus our discussion. "Do you think Mrs. Jones did the right thing?" I ask. "Try to support your answer with some examples from the story."

"No, she's dumb."

"Hey, I thought it was nice. I would've done the same thing."

As the conversation gets going again—with no regard to our discussion guidelines—I notice Alana and Jasmine talking to each other. Suddenly there are about six different side conversations going. I strain to hear how many of them are focused on the text. None, to my dismay. The students get louder and sillier until I interject once more. Our conversation, if it can be called that, continues like this until I end it, thinking we need to do something about our discussion habits, and do something quickly.

After class I looked over the middle school speaking and listening standards and wondered what we could do to meet the group meeting standard, E3b of the *New York City Performance Standards*:

### Speaking and Listening Standard E3b

The student participates in group meetings in which the student:

- displays appropriate turn-taking behaviors;
- actively solicits another person's comment or opinion;
- offers own opinion forcefully without dominating;
- responds appropriately to comments and questions;
- volunteers contributions and responds when directly solicited by teacher or discussion leader;
- gives reasons in support of opinions expressed;
- clarifies, illustrates, or expands on a response when asked to do so; asks classmates for similar expansions;
- employs a group decision-making technique such as brainstorming or problem-solving sequence.

As I read the standard, I thought back to steps I had taken before to improve my own teaching. The use of action research during the previous school year had helped greatly. The Education Resources Information Center (ERIC) defines action research as "research designed to yield practical results that are immediately applicable to a specific situation or problem." More simply put, action research is a way of saying, "Let's study what's happening at our school and in our classroom and decide how to make it a better place (Calhoun 1994, 19)."

I was introduced to the idea of action research through the National Teachers Policy Institute (NTPI). I am a MetLife Fellow in NTPI, an organization dedicated to bringing teachers' voices to policy making. We use action research to help us become more reflective practitioners and as a forum for discussing education policy. During the 1998–99 school year I researched how my struggling readers used independent reading time during class. I discovered from my data that these students were rarely on task and were choosing inappropriate texts to read. Based on this information, I ordered more engaging books and reorganized my library to promote texts that were at their reading level so that they would be more focused. One of the policy implications that we discussed at NTPI was that teachers need more discretion in ordering resources. Having had success as a teacher using action research, I wondered what would happen if the students conducted their own research on their discussion habits.

I designed a research form, a variation of the "field notes" from Hubbard and Powers' helpful book *The Art of Classroom Inquiry* (1993). Trying to keep the research simple for the first time, the observer of our discussions would note three things: (1) T = when a student spoke out of turn, (2) N = when a student had a side conversation with a neighbor, and (3) R = when a student raised his or her hand. Although my long-term goal was for the class to have an authentic conversation without raising hands, there had been too many interruptions the day before. We needed to begin by using a basic method of taking turns. I invited the students' science teacher to collect data during our next discussion to model the process for them.

At our next session, students continued to speak over one another and to conduct side conversations. The discussion again veered from one topic to the next. Shown here is a graph of the data from this first day of research. Although these data are an approximation, the results accurately depict how our conversations went during these first days of school:

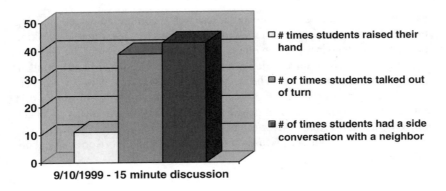

50
40
30
20
10
0

□ # times students raised their hand

▣ # of times students talked out of turn

▣ # of times students had a side conversation with a neighbor

**9/10/1999 - 15 minute discussion**

I did not know what to expect when I presented the research to the class, but I guessed that the students might be defensive about the results. I explained how we could use research to help us solve the problems we face as a community of learners, then I unveiled the research results. The shocked look on many of their faces confirmed the impact this type of classroom research could have on student work. Many couldn't believe that this was how our discussion from the day before looked. Although some did want to argue with the data, the conversation quickly turned to the question of what to do about it. The responsibility was on the class to offer strategies for improvement. We agreed that most of us would learn more in relocated seats (e.g., not next to their friends) and that we needed to speak one at a time, which meant raising our hands for now.

The impact of the research was evident in a conversation between Adriana and Kayla. The day before, they had been "neighbors" in the meeting area and according to the research had side conversations seventeen times! At the beginning of our next class, Kayla came in to sit down next to Adriana. Instead of chatting with Kayla as usual, Adriana said, "You know what, you probably shouldn't sit there. You know we're going to talk if you do." Sure enough, Kayla moved and they did a much better job than the day before.

For the next three weeks student researchers collected data from our fifteen- to twenty-minute discussions, and regular conversations were held to explore how to improve our discussions. It was exciting for everyone to see the progress we made over that time:

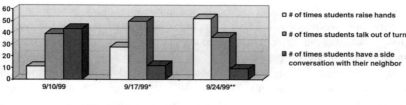

\* majority of talking out done by two students
\*\* talking out of turn was more focused on discussion

# A Class of Researchers

Because of my success with this initial research, I committed myself to using the process to improve discussions throughout the year. Meeting the speaking and listening standards became one of my instructional goals for the year. We investigated our class conversations, our work with book partners, and our talk in peer writing response groups. The students continually worked together to meet new challenges as they arose, and I used our analysis to guide my instruction. At times the process became too much of a routine or was seen as a way to "check up" on students' actions. Most often, though, it greatly helped our efforts to become smarter about our discussion habits and to think more critically about our reading and writing.

By March, students were working together in book clubs. Three to five students read the same book and had book club meetings during class. When planning how to run the book clubs, I considered assigning roles so that they knew how to participate in the meeting; but although roles can be a helpful start to a book club, they often limit participation. Kids focus only on their role of "word wizard" or "super summarizer" and then don't have real conversations about the book and their own ideas. I ultimately decided it would be more dynamic to work on having real-life conversations about books.

During the book partnerships the month before, I had tried a technique known as a *fishbowl* to model good book talks. During a fishbowl, certain students (the fish) do their work in the center of the room (the bowl) for the rest of the class to see. Knowing the difficulty we faced implementing the fishbowls, a professional developer with whom I collaborated on a regular basis suggested visiting another school in the district in which exciting work was being done with fishbowls and classroom research. Having this opportunity to talk with fellow teachers who were involved in the same kind of work was an invaluable experience. I returned to school ready to embrace the challenge of this new form of investigation.

Knowing that one essential element for any good conversation is an interesting topic, I chose a thought-provoking book for the class to read. *I Hadn't Meant to Tell You This*, by Jacqueline Woodson, is the story of Marie, a well-off young African American girl, and her relationship with Lena, a poor white girl who is labeled as "white trash" at their predominantly African American school. Lena and Marie have both lost their mother, and Lena is being sexually abused by her father. The challenges these two girls face in life and in their friendship spoke to the children in my class. After each day of reading, the students would either be fishes in the inside circle or researchers outside the circle. They would work with partners so that while one was in the discussion the other was observing and recording data. Afterward, the students met to discuss the research and set goals for the next book talk.

The conversations that took place in March bore little resemblance to our work in September and October. The biggest change was that the students were responsible for the whole discussion, from their actions as participants to the content of the conversation. They began by proposing discussion topics based on the reading. Once I had recorded those on the board, the students spent the next ten to fifteen minutes discussing the topics among themselves.

In September and October we had looked at simple actions such as raising our hands or talking out. Now we were investigating the specific content of the discussion as well as more sophisticated behaviors such as body language and eye contact. Our new research form resembled in many ways the guidelines for the Speaking and Listening standards (see Figure 12–1). (See Jody Madell's entry in the Samples section for another example of an "accountable talk" classroom research form.)

# RESEARCHING CLASS DISCUSSIONS

Researcher _____ Fish _____ Date _____

We try to use all of these *discussion strategies* to have a good book talk:

| Strategy | Evidence (Researcher: Check off each time the strategy is used by your partner) |
|---|---|
| Proposes a discussion topic | |
| Shares ideas about a topic | |
| Builds on another person's idea | |
| Makes wise connections | |
| Uses text to support ideas | |

We try to *do* the following things to have a good book club meeting.
Researcher: Note how your partner did with the following actions.

1 = little or no evidence of action
2 = some evidence of action
3 = much evidence of action

| Action | Evidence and Comments |
|---|---|
| Tries not to dominate | |
| Takes turns speaking | |
| Has good body language and eye contact | |
| Makes positive comments | |
| Pulls others into the discussion | |

Researcher's Feedback _____
_____

Fish's Goal from Previous Discussion _____
_____

Circle one:              Still working on goal
                         Almost at goal
                         Met goal

Fish's Goal for Next Discussion _____
_____

**Figure 12–1.**

# Getting Smarter

As our research continued we began to notice patterns in both process and content. Student researchers became adept at focusing on subtle clues to help them better understand discussion habits. Maureen, for instance, noticed that her partner Jena nodded after comments several times, wanting to say something, but wasn't given the chance to speak. Tyrell suggested that it was the individual's responsibility to jump into the conversation. After further debate we agreed that it isn't always that easy, especially when there are different personalities in the class as well as students from diverse cultural backgrounds. For some, it might not be the norm to speak so freely and jump into a conversation.

We decided to allow for some pauses during the book talk to give less vocal students the opening to share ideas. We also pointed out that asking reticent students point blank, "What do you think?" to pull them into a conversation made it difficult for them to respond. Instead, we tried to use more inviting language, such as "Do you agree with what Maureen said?" or "What do you think about this character's actions?" Ultimately, we set a goal that during each discussion each participant would share at least one meaningful idea. My next lesson focused on the difference between just saying something and sharing a meaningful idea. Students would respond to a question by saying, "Yes, I agree," without elaborating. We worked on how we might elaborate our ideas and support our opinions. Soon, the class was going back into the book to clear up confusion or to make a particular point.

I documented our progress in meeting our goal of having everyone meaningfully involved in the conversation. Over our last six discussions, we were able to meet our goal twice. There is still much work to be done, but on the days we met our goal the class experienced a sense of accomplishment:

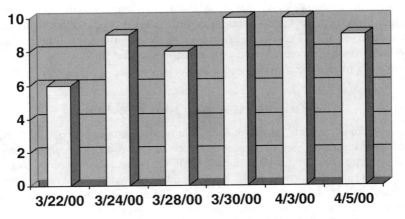

☐ Number of students out of the 10 in the fishbowl who shared
at least one meaningful idea during their ten to fifteen minute
book talk.

Not only did we work to meet class goals, but students also set personal goals for their book talks. Juan's partner noticed that he was always the first one to answer a question. Juan made it his goal to let others answer first and listen to their ideas. During his next discussion, Juan worked to meet his goal, and his partner noted his progress. Juan found that as soon as he was aware of his habit of speaking first, it was easy to curb. He also realized how much more there was to learn when his own ideas didn't always dominate the discussion.

In addition, our research helped my instruction during our read-alouds. I noticed that students were interested in discussing the characters from *I Hadn't Meant To Tell You This*, but sometimes they lacked the questioning strategies to expound on their ideas. This was also the case when they had their conversations about their book-club books. My lessons during read-alouds helped them focus on asking open-ended questions about a character's motivation and then using the text and the events of the story to support their ideas.

## Students' Reactions to the Research Process

In response to reflective questions about the research process, the class agreed that doing the research positively affected both the class' discussion habits and students' individual roles in discussions. Tyrell, a very dominant speaker, claimed that the research process helped him "because I saw how other people can't talk. Being outside [the discussion] makes you see not to talk so much." In turn, Maureen noted that our fishbowls also "made the shy people speak more." Zheng, one of the "shy people," agreed that "doing the research made me want to talk more. I knew it was important to share my ideas." A significant number were also able to make the connection between researching and setting goals for our class discussions and their work in the book club meetings. Lorena shared that the fishbowl helped her "learn how to express my feelings about the book more in our book club meetings." Finally, Adriana noted that the research "helped me because I got a different view from other people. I know that now they can see what I can't see." Adriana thus eloquently defined the research process that helped our whole class get a different view and become smarter in our work toward the speaking and listening standards.

## Putting Research into Practice

For students to meet "new" standards that require problem solving and critical thinking, "new" learning must occur in the classroom. In their discussion of standards-based reform in education, Thompson and Zeuli contend that "in order to learn the sorts of things envisioned by reformers, students must think (1999, 8)." By "think" they mean that "students must actively try to solve problems, to resolve dissonance between the way they initially understand a phenomenon and new evidence that challenges that understanding, to put collections of facts or observations together into patterns, to make and test conjectures, to build lines of reasoning about why claims are or are not true (9)." Although

Thompson and Zeuli focus their discussion of reform on math and science, their definition of the thinking process accurately describes the work our class did to meet the speaking and listening standards.

Thompson and Zeuli raise the question, "What features must characterize the pedagogy of professional development if it is to help teachers learn [how to have students 'think' in their classroom] (21)?" They argue that to address this question, "it seems most productive to begin with an image of the teacher as a learner (21)." The use of action research has enabled me to be a learner in the classroom and to look critically at my practice as I struggle to help my students learn and meet the rigorous new standards. In turn, the action research has helped my students "think" about how we are going to meet the standards and become more intelligent about our class discussions.

Professional development in a school or in a district can be structured to empower this reflective teaching and learning. For classroom research to happen, it is essential that teachers:

- *develop their own research questions based on the students' needs, the teachers' passions, and the learning standards the students are expected to meet*
  Action research questions shouldn't be mandated from outside but should grow from classroom situations, critical conversations between colleagues, and the guidance of professional developers and instructional leaders in the building. (This is how I developed my original question, What would happen if the students researched their discussion habits?)

- *have the time and the tools in class to pursue their research*
  This kind of classroom research is effective only if it is integrated into the day-to-day work of the teachers and the students.

- *collaborate with other teacher researchers and collectively analyze the data and reflect on their ideas*
  Releasing teachers to observe one another's classrooms and providing time to meet with a professional developer are two such examples that contributed to my success with classroom research this year.

For the students, it is important that:

- *the research is recognized as critical thinking essential to meeting the new standards*

- *the research is done in a safe environment and in the spirit of problem solving*
  Students are willing to take risks and be honest about their work if that honesty is not used against them in the form of grades or disciplinary measures.

- *students are purposefully collecting data*

- *students, just like teachers, are given the opportunity to collaborate with one another and share their ideas about the data*

When research became a classroom routine we grew into a strong learning community, one in which students could share honest concerns and criticisms knowing they would be handled in a proactive manner. Although the instructional focus of much of our work was on discussion and meeting the speaking and listening standards, we learned much more than better discussion habits. The research process enabled the students to view their learning habits in new ways, making them better learners. In the words of the Institute for Learning, I have "called on children to use the skills of intelligent thinking, and by holding them responsible for doing so [I am teaching] intelligence (1999)."

## Works Cited

Calhoun, E. F. 1994. *How to Use Action Research in the Self-Renewing School.* Alexandria, VA: Association for Supervision and Curriculum Development.

Hubbard, R., and B. Powers. 1993 *The Art of Classroom Inquiry.* Portsmouth, NH: Heinemann.

Institute for Learning. 1999. *The Principles of Learning.* Pittsburgh: Learning Research and Development Center at the University of Pittsburgh.

Thompson, C. L., and J. S. Zeuli. 1997. *The Frame and the Tapestry: Standards-Based Reform and Professional Development.* Ann Arbor: Michigan State University.

# Bridge

In designing our book I had to decide whether material on engagement fell within the scope of our theme. I concluded that it did, that it allowed us to pull back from our work in the classroom to measure it and to see with the greater objectivity that math can provide. Now I see another way it addresses our theme: children get the inadvertent benefit of seeing how math looks when it is applied (which should please David Hardy and other math modelers). They form a clearer understanding of percentages, bar graphs, ratios, and the like after seeing their own behavior represented. Thoreau said, "The more you have thought and written on a given theme, the more you can still write. Thought breeds thought. It grows under your hands (Thoreau 2001, 11)." So it is here.

An important way our thoughts have grown under Matt's hands is in his focus on action research, which will produce the data teachers need to grow in their work. The data should convince anyone, but before teachers will be willing to do the math they must be willing to embrace the idea that, as Matt puts it, "all students can develop intelligent habits of mind." If this book does anything right, it will contribute to the changing of beliefs to admit this possibility.

I can't help but note here that one of the prime objectives of much literary fiction is to demonstrate how characters change their beliefs about people and situations. If the characters don't change their beliefs, it is likely that the author is setting out to change the reader's beliefs. As Diane Lefer, who has contributed a literary sample to this book puts it, "if there has to be a turning point, why not make a shift in the *reader*? What if the *reader* changes and comes to see a character or a situation in a new light (Lefer 1994, 18)?" Our intent is to change readers with *this* story. Perhaps if we can't persuade them, readers might conduct their own research, see for themselves.

The fact is, we all need to build better "habits of mind," which should improve all of our chances to succeed, just as civilization itself began to triumph over the so-called Dark Ages when, in the Renaissance, the question of scholars changed from Why?—which took them to answers based on religious ideas—to How much? Seeing our behavior represented with some degree of objectivity calms us, makes us more reasonable, improves our judgment, makes us fairer. It levers our egos out of the way, relieving us of having to defend them or to suffer because of them. It gives shy people a chance.

Matthew Szenher, who contributed earlier to our collaborative chapter on poetry in the math classroom, illustrates in the following chapter another way to give shy people a chance: through apprenticeship.

# Works Cited

Lefer, D. 1994. "Breaking the 'Rules' of Story Structure." In *The Best Writing on Writing*. Cincinnati: Story Press.

Thoreau, H. D. 2001. "Illuminations: Great Writers on Writing." *Teachers & Writers* 33 (1): 11.

# 13

## The Mathematician's Apprentice

### Matthew Szenher

I couldn't hit a backhand for the first five years of my tennis-playing life—five years of lessons, practice, and matches. Finally, watching me hit one atrocious backhand after another, a high school coach pulled me onto a court with a ball machine on the other end. He stood behind me and we hit backhands together as one person. He let me make the motion until he felt it go wrong, at which point he would take over and show my brain when to react and my muscles what to do. After fifteen minutes of this, he let me hit the backhand by myself while he watched. The shots were much improved though far from perfect. I remained at this plateau for a few weeks. Then the backhand clicked. I could now hit high balls, low balls, balls with slice and balls with topspin—all with great precision.

The evolution of my backhand came to mind when I took on a special project with a student named Lawrence Cinamon, who had been having as bad a time with mathematics as I had had with tennis. Lawrence had failed a traditional, learn-by-repetition ninth-grade math course and hadn't done much better in tenth grade in another school. He transferred to my school (the Dwight School, in New York City) in the tenth grade and entered my International Baccalaureate Computer Science class the same year. Although the class is normally open only to juniors and seniors, Lawrence had performed precociously on certain entrance exams, qualifying him to enter the class one year early.

Precocious or not, his early impact on the class was more a ripple than a tidal wave despite his red hair, which I couldn't help but associate with the spice whose name is pronounced like his and carrots, a food that I love. He was naturally shy, and afflicted with a stutter that got worse when he was uncomfortable. His assimilation into class was not a simple process. His classmates helped, however. There were only three other students, all boys, and the small number lessened the fear of speaking. They were kind to him and made fun of him only gently. (High school males *must* rib one another.) For my part, I let Lawrence be himself. One day a few months into the school year he drew a rendition of Mr. Peanut on the whiteboard with monocle, top hat, and cane, taking care to get the contours of the dimpled body just right. Lawrence couldn't reach high enough to finish the top hat, so I did. Even though this intruded on class time (only forty-five minutes), I let Lawrence draw it then and many more times after. I felt it was important to encourage Lawrence's eccentricities.

One of Lawrence's eccentricities was his tendency to make insightful remarks that had absolutely nothing to do with what was going on around him. This, too, I encouraged, if only by not discouraging it. These non sequiturs came abruptly. "Mr. Z!" he would say, in the middle of a lecture about the evolution of computer architecture. Then he would proceed to offer some elaborate theory on the failure of the French to defend themselves in World War II, or question the prevailing wisdom about use of the price-to-earnings ratio in the valuation of companies. One day he sidetracked the class once too often, so I suggested that he write his tangential thoughts down in the back of his notebook. I promised to discuss them later, when time permitted.

At the end of the course Lawrence's classmates graduated in May, but Lawrence, a junior, wouldn't leave for summer break until June. This happy accident left him with no agenda for two weeks, forty-five minutes a day. Not wanting "to sit idly by like a sloth," as he put it, he agreed to undertake a mathematical investigation. I suggested he do this for two reasons: to determine if he could benefit from a different approach to mathematics education than he'd experienced before, and to help me in my struggle to incorporate into my mathematics classroom constructivist methods—to which I had recently been introduced. (I teach IB Mathematics as well as IB Computer Science at my school.) I was searching for a way to marry the rather open-ended constructivist approach to teaching with the skills-based mathematics education emphasized by my department chair. We had to go only as far as the back of Lawrence's math notebook for the topic of his investigation. One of the questions he had written there was, "Is a base $\pi$ number system feasible?" In other words, "Can a representational number system be constructed on the quantity $\pi$ just as the decimal system is built on the quantity ten and the octal system on eight?

Here is the story of Lawrence's research in his own words, followed by his appendix on positional number systems:

> On the first day of our research, Mr. Szenher and I explored the possibility of base $\pi$ representations of numbers. We immediately ran into a problem. Base ten numbers use digits 0–9 and base eight numbers 0–7; there is an obvious pattern here. But which digits should be used for base $\pi$ numbers? We tried to use digits 0–3 but found it impossible to represent certain quantities (like 2!). Other choices of digits also failed.
>
> Following Mr. Szenher's suggestion, we turned our attention temporarily to the seemingly simpler possibility of number systems with fractional bases. We first looked at the representation of different quantities in bases $\frac{1}{2}$ and $\frac{1}{3}$. We tried using the base three digits 0, 1, and 2 for our base $\frac{1}{3}$ representations.

We found that a quantity in base three is represented as the reverse string of digits in base $\frac{1}{3}$. For example:

$$1.012_3 = 210.1_{\frac{1}{3}}$$

The same equality property holds for bases two and $\frac{1}{2}$ and in fact any bases $x$ and $\frac{1}{x}$. This discovery was interesting, but did not lead us very far. Representations in other fractional bases like $\frac{2}{3}$ or irrational bases such as $\pi$ remained a mystery. Mr. Szenher and I left for our weekend break somewhat frustrated.

On Sunday, I was sitting on my terrace and I came up with the idea to write a formula that would convert a number in base $x$ (where $x$ is an integer greater than 10) to its decimal value. I came back to school on Monday and Mr. Szenher and I, to get a better sense of the problem, graphed the decimal values of the base ten, base eleven, and base twelve numbers 1–30 (see Figure 13–1): We omitted representations with non-decimal digits (like $1A_{12}$) as these would have left gaps in the graph. I immediately noticed that at every multiple of 10 on the $x$ axis, there was a jump in the graph. I took special note of this and it

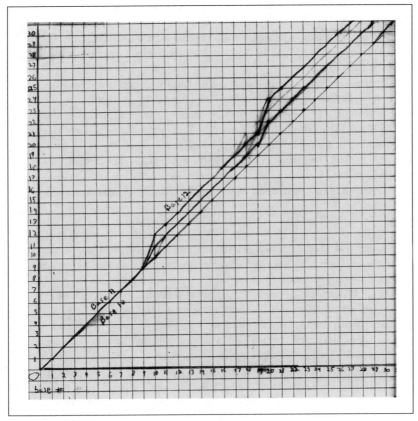

**Figure 13–1.**

remained at the forefront of my thought for the day. Later that evening, with the day's discovery still in mind and again sitting on my terrace, I came up with the following conversion formula:

$$(x - 10) \left( \frac{base_x \ value}{10} \right) + base_x \ value = base_{10} \ value$$

where $x$ is the base of the number system converted; all arithmetic is performed in base ten. That evening, Mr. Szenher, knowing that I was working on a conversion formula, came up with a computer program to do a "brute force" conversion from base $x$ to base ten (as in Figure 13–1) that we could use to check my conversion formula. The program calculated, individually, the base ten representation of all the numbers from 1–10 digits long and containing only digits in the range 0–9 inclusive in bases ten to twenty.

The next day, we came to class excited to share the fruits of the previous night's labor and to compare results, my formula and his program. We found that the formula did not produce the same numbers that were generated by Mr. Szenher's program. For instance, when trying to find the decimal value of $13_{12}$ (i.e., 13 in base twelve), my formula yielded $15.6_{10}$. Mr. Szenher's program generated $15_{10}$, which was correct. In fact, my formula only worked when the base $x$ representation was a multiple of ten when interpreted as a decimal number (i.e., $20_{11}, 40_{12}$, and so forth). Naturally, I was quite perturbed by this turn of events.

However, Mr. Szenher had an ace up his sleeve. Seeing the error in my formula, he told me about a function called "floor." Floor (signified by $\lfloor \ \rfloor$) is used to round a real number down to the nearest smaller integer. For example, $\lfloor 13.65 \rfloor$ is 13. I figured out how I could modify my conversion formula to take advantage of this function, realizing that the problem arose with the fraction that was created by the operation $\frac{base_x \ value}{10}$. I saw that once that fraction was eliminated, the formula would always produce a whole number for an answer. With the addition of the floor function, my conversion formula became:

$$(x - 10) \left\lfloor \frac{base_x \ value}{10} \right\rfloor + base_x \ value = base_{10} \ value$$

where $x$ is defined as before. We suspect that this formula works for every number represented in a positional number with base ten or more that contains digits only in the range 0–9 (i.e., the formula will not convert the number $AF6_{16}$). We have no conclusive proof of the formula's validity, however the formula did work for every number that we tried. Mr. Szenher believed that a proof was beyond our current abilities.

A few days later, after school, I walked home on Broadway as usual. The weather was perfect, and I used the opportunity to reflect on this project. I was between Eighty-seventh and Eighty-sixth when the following question popped into my head: "How useful is the formula?" After all, the formula does not work with base $x$ numbers that contain digits outside the range 0–9. On the other hand, the time it takes to convert a number from base $x$ to base ten does not depend on the length of the number, as is the case with traditional conversion methods (see Figure [13–1]). Only a constant number of arithmetic operations (one subtraction, one division, one multiplication, and one addition) are required. I walked past the Gap on the corner of Eighty-sixth and began crossing the street. In the middle of the street I realized how I could quantify the utility of the formula. I call this the Utility Function:

> Probability of being able to use the conversion formula (i.e., that all digits are between 0 and 9, inclusive) $= (\frac{10}{x})^n$ where $x$ is defined as before and $n$ is the number of digits in the number to be converted.

I must have been walking very slowly or standing still by this point because I was being honked at by motorists who must have been trying to get out to the Hamptons in a hurry. I was so absorbed that I thought of the exact function in less than a minute. I was completely unaware of my surroundings. I guess that's exactly the state that Archimedes was in after he displaced water from his bathtub. There were only two significant differences. I was on Broadway and I had most of my clothes on. As soon as I got home I wrote the formula in my notebook.

The Utility Function allows us to see what the probability is of being able to use the formula given the amount of digits and the number system of the non-decimal number. I returned to school on Monday, excited to show Szenher my Utility Function. He asked me to write a proof for the function and I was able to do so (with a little help from Mr. Szenher) using my knowledge of probability and of mathematical proofs, which I had obtained in tenth-grade:

> Suppose we are trying to use the formula to convert an $n$-digit number $a_1 a_2 \ldots a_{n-1} a_n$ that is represented in base $x$, $x > 10$. In order to use the Utility Function to convert $a_1 a_2 \ldots a_{n-1} a_n$ to base ten, every digit $a_i$ must be between 0 and 9. The probability that $a_1$ is between 0 and 9, inclusive, is $\frac{10}{x}$. So, too, the probability that $a_2$ is between 0 and 9 is $\frac{10}{x}$. In fact the probability that any digit $a_i$ is between 0 and 9 is $\frac{10}{x}$. So, by the counting principle, the probability that every digit in the $n$-digit number to be converted is between 0 and 9, inclusive, is $(\frac{10}{x})^n$.

Mr. Szenher plotted the results of the Utility Function using Microsoft Excel (see Figure 13–2). One can see from this graph that the probability that our conversion formula can be used to convert a base ten number to the same base ten number is always 1. For bases larger than ten, the probability decreases exponentially as digits increase.

## Appendix
### Positional Number Systems and Conversion
### to Decimal from Other Bases

A positional number system provides a method for compact representation of real numbers. Some familiar examples of positional number systems are decimal and binary. Most positional number systems are associated with a quantity known as a base or radix, $r$. The base $r$ of the decimal system is ten and the binary base is two. The value of $r$ can be any integer greater than 1.

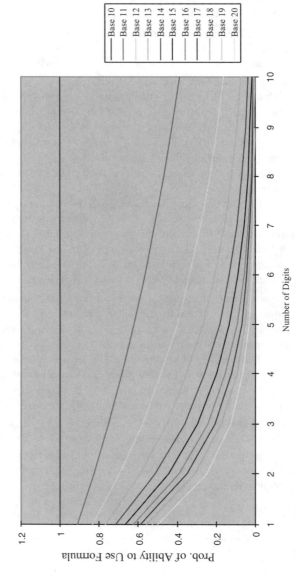

Probability of Ability to Use Formula For Bases 10-20

**Figure 13–2.**

A real number is represented in a given number system as follows:

$$a_n a_{n-1} \ldots a_1 a_0 . a_{-1} a_{-2} \ldots a_m$$

where each $a_i$, $m \leq i \leq n$, is an integer and $0 \leq a_i \leq r - 1$. For the representation $452.5_{10}$ (i.e., four-hundred fifty-two point five in base ten):

1.    $r = 10$
2.    $n = 2$
3.    $m = -1$
4.    $a_2 = 4$
5.    $a_1 = 5$
6.    $a_0 = 2$
7.    $a_{-1} = 5$

The representation $a_n a_{n-1} \ldots a_1 a_0 . a_{-1} a_{-2} \ldots a_m$ is shorthand for the following sum of products:

$$a_n r^n + a_{n-1} r^{n-1} + \ldots + a_0 r^0 + a_{-1} r^{-1} + \ldots + a_m r^m.$$

Therefore $452.5_{10}$ represents the sum:

$$4 \times 10^2 + 5 \times 10^1 + 2 \times 10^0 + 5 \times 10^{-1}$$

or, in other words, four 100s, five 10s, two 1s, and five .1s. It turns out that any real number can be represented by some string of values in any base.

If the base $r$ is greater than ten, then more than the ten digits 0–9 are required to represent certain real numbers. For example, base sixteen requires 16 digits (properly called hexadecits). These hexadecits are 0–9, A (which represents 10), B (which represents 11), C (which represents 12), D (which represents 13), E (which represents 14), and F (which represents 15).

To convert the base $x$ representation of a real number to the base ten representation of the same number, simply expand the base $x$ representation to its sum of products form and perform base ten arithmetic on that sum. For example, to determine the decimal representation of $10201.2_3$:

$$10201.2_3 = 1 \times 3^4 + 0 \times 3^3 + 2 \times 3^2 + 0 \times 3^1 + 1 \times 3^0 + 2 \times 3^{-1}$$

$$= 81 + 0 + 18 + 0 + 1 + \frac{2}{3} = 100.6\bar{6}_{10}$$

To determine the decimal representation of $AF1_{16}$:

$$AF1_{16} = 10 \times 16^2 + 15 \times 16^1 + 1 \times 16^0 = 2800_{10}$$

The process that Lawrence describes went far beyond anything Lawrence had yet encountered in his traditional mathematics classes. It was an apprenticeship, with all the nuance and intimacy of the apprenticeship I had undergone years earlier to learn to hit a backhand. Moreover, Lawrence's apprenticeship accomplished at the deepest level what constructivist educators and theorists recommend.

For many students mathematics is as obscure and otherworldly as magic. Constructivists argue that this attitude is a symptom of our traditional pedagogy:

"The traditional teaching method of teacher as sole information-giver to passive students appears [to produce students who] are unable to successfully integrate or contrast memorized facts and formulae with real-life applications outside the school room (Hanley 1994)." This was true of Lawrence, who once commented, "We [math students] had to memorize very formal proofs but we didn't actually make them our own. We just had to regurgitate stuff from the books." He said this method wasn't "productive [or] interesting."

According to J. G. and M. G. Brooks (1993, 25) a constructivist teacher should:

- Become one of many resources that the student may learn from, not the primary source of information.
- Engage students in experiences that challenge previous conceptions of their existing knowledge.
- Allow student responses to drive lessons and seek elaboration of students' initial responses. Allow student some thinking time after posing questions.
- Encourage the spirit of questioning by asking thoughtful, open-ended questions. Encourage thoughtful discussion among students.
- Use cognitive terminology such as *classify*, *analyze*, and *create* when framing tasks.
- Encourage and accept student autonomy and initiative. Be willing to let go of classroom control.
- Use raw data and primary sources, along with manipulative, interactive physical materials.
- Don't separate knowing from the process of finding out.
- Insist on clear expression from students. When students can communicate their understanding, then they have truly learned.

Many of these criteria are evident in Lawrence's story. He had significant autonomy, spending weekends devising solutions and formulating new questions without me. He generated raw data using a program that I wrote and generalized that data to create formulas. He provided various expressions of his work: the story he wrote, the proofs it contains, and my interview with him afterward. We both posed open-ended questions that led to deep and interesting investigations, and more questions.

That the experience was important to Lawrence is clear from the interview I conducted with him after our work together:

*Szenher:* How would you describe your apprenticeship?

*Cinamon:* It was very productive and interesting.

*S:* How so?

*C:* Because . . . I collaborated with a teacher instead of a student.

*S:* How is that different?

*C:* The teacher knows what he's doing.

*S:* [laughs] Well, we want you to think so.

*C:* There's more insight, more experience.

*S:* Insight into what?

*C:* For instance, I or another student wouldn't be able to come up with the floor function or how to write a proof.

*S:* You found this [apprenticeship] fun?

*C:* A lot of fun.

*S:* Do you think in the future, you'll do more [mathematical] explorations?

*C:* I think so.

Isn't this what we wish of all our students: that they have fun with math, learn at the same time, and be inspired to continue?

## Works Cited

Brooks, J. G., and M. G., Brooks. 1993. *In Search of Understanding: The Case for Constructivist Classrooms*. Alexandria, VA: Association for Supervision and Curriculum Development. In "On Constructivism," S. Hanley. Maryland Collaborative for Teacher Preparation.

Hanley, S. 1994. "On Constructivism." Maryland Collaborative for Teacher Preparation.

# Bridge

If *Teaching for Depth* has a plot line, Cinamon's essay is arguably at the summit in the way it realizes our thesis so *pleasurably*. (One of the central aims of this book is precisely that: to recover the lost pleasures of mathematics.) I point with a heavy note of justification to the way Cinamon's pleasure is made evident: through the telling of a story, that ultimate expression of the humanities.

One of the key elements in storytelling is the use of voice. A voice has a tone. Both Szenher and Cinamon capture their pleasure in this tone: "despite his red hair, which I couldn't help but associate with the spice whose name he shares, and carrots, a food that I love." Szenher is inviting us to share his senses to give us insight into his avuncular relationship to Cinamon. Cinamon himself writes about coming up with ideas on his terrace, which *sets* the tone, as it were: "Later that evening, with the day's discovery still in mind and again sitting on my terrace, I came up with the following conversion formula. . . ." Then he tells of the revelation on Broadway: "I must have been walking very slowly or standing still by this point because I was being honked at by motorists who must have been trying to get out to the Hamptons in a hurry. . . ." History assists him in conveying his pleasure as he continues, "I was so absorbed that I thought of the exact function in less than a minute. I was completely unaware of my surroundings. I guess that's exactly the state that Archimedes was in after he displaced water from his bathtub. There were only two significant differences. I was on Broadway and I had most of my clothes on. . . ."

So what's the root of this pleasure, conveyed in the tone? My own hypothesis is that pleasure doesn't need a cause. It is already there in the sensation of living, with mathematical intuition and cogitation being an inherent part of life. It needs nothing more than to be allowed to exist. Several factors went into the permission in this case: a supportive and sensitive teacher, a motivated student, and time. I find it significant that the time actually came outside the curriculum. Nothing was there to impede Cinamon's questions. In the conclusion I will talk about how I often learned more in extracurricular activities than elsewhere. Indeed, because of the recent death of my mother I was at our family home in Baton Rouge looking through old collections of papers, and came across my report cards. Ouch. I was earning abysmal grades in boringly designed classes while the newspaper I edited was winning awards. I identify with Cinamon. Would that Matthew had been my teacher.

The question becomes, How do you take this formula for maximum engagement and convert it for use in a classroom full of not one, but twenty, thirty, sometimes forty individuals? My answer is to work with student portfolios. Portfolio assessment has been researched and described in great detail elsewhere, but I include something on it for its relevance to engagement. It is at the center of my work as a staff developer because of its potential to empower students, to allow them to work at their own pace, to ask their own questions, to demonstrate their learning in a variety of ways, to deepen their inquiry until they understand concepts thoroughly . . . the list goes on.

When working with teachers I have often resorted to drawing a simple pattern to help illustrate its power to deliver a sense of meaning to the process of learning:

The conventional (seemingly endless) lineup of lessons and tests looks like this in my schema:

-/-/-/-/-/-/-/-/-/-/-/-/-/-/-/-/-/-/-/-/-/-/-/-/-/-/-/-...

Each dash is an isolated lesson or activity. Each slash is a test. You might call it the *slash-and-dash* method. Kay Rothman points out that the slash-and-dash method fosters the "Zeigarnick effect," i.e., "if you finish the task, it's out of your mind." "That's why people forget what they learn for tests. You learned it; you used the learning on the test; now you don't need it any more." The slash-and-dash method of teaching is not a pleasant one. It is the one used to teach me in most of high school and most of college. It is life lived in "chunklets" (a word I picked up on the Web, checking out what I could find on the Zeigarnick effect). It is dry cat food compared to the gourmet cooking of enlightened education. I reject it.

Well-designed portfolio teaching looks like a spiral, in my schema. One's learning is a recursive path. The dry chunklets become juicy chunks pleasurably and purposefully digested into the whole, which is continuously growing as a whole. This is the method I use for my own learning, even outside of school. I visit ideas and return to them later with greater depth of understanding. Or I try to. I have certainly tried to develop this book according to the pattern.

A friend from Expeditionary Learning Outward Bound (ELOB) once passed along to me that organization's handy mnemonic for portfolio process: "Collection—Selection—Reflection—Projection." (At least I think it originates with ELOB. I await correction.) I use it sometimes to capture the idea in a nutshell. You collect work in an archive, take time at an appropriate point in the curriculum to pick out some of the work according to specified criteria ("best work," "most challenging work," "work that needs improvement," "work that demonstrates certain understandings or skills," etc.), revise and reflect on the work and on your learning process, then project or predict how your work and your learning processes will look in the future. Portfolios may be used for many purposes, of course. Some are "passage portfolios," for instance, and others are "process portfolios." But in all cases they inevitably serve as engines for continuity in learning.

David Berlinski discusses mathematical concepts of continuity in *A Tour of the Calculus*: "Continuity is an aspect of things as rooted in reality as the fact that material objects occupy space; it is the contrast between the continuous and the discrete that is the great generating engine by which the real numbers are constructed and the calculus created. The concept of continuity is, like so many profound concepts, both simple and elusive, elementary and divinely enigmatic (1995, 130)." I borrow his words to describe continuity in learning: it is a profound concept, both simple and elusive, elementary and divinely enigmatic.

Portfolio design deserves far more discussion than is remotely possible in this context. My own notes on the work being done in the Lab School could fill a volume. In fact, I have conducted weekly "Portfolio Lunches" at the school for several years now, where teachers voluntarily come to share their ideas. We discuss teacher portfolios as well. Teacher portfolios are seen as a way to tell the story of our discoveries as researchers. We establish themes and essential questions, gather evidence to support our hypotheses, challenge one another's thinking. I, for instance, was taken heavily to task for coming in with too vague a thesis for my portfolio. I later sheepishly revised it. Over the year I have gradually explored it, mulled over it, gathered material to support and refute it. The theme is centered around the idea of trust, that "nothing of worth can be accomplished in education without trust." I suspect I will be elaborating and revising the theme for years. It is, however, already working for me, just as the theme of this book helps keep it focused and organized. When I encounter a lack of trust in my work, rather than react with the unhappy feelings that can usually flourish as mistrust's boon companions, I study its nature and its dynamics. This reflective stance is always difficult for me to achieve but it gives me at least a fighting chance to strive for understanding.

I hope that my brief pass through the subject of portfolios has adequately conveyed the importance I feel they have in the classroom and, whether formalized or informalized, in life. In their structure they offer the possibility of a kind of coherence and meaning that is virtually impossible in traditional education and in our slash-and-dash consumerist culture that forbids reflection. It's in the math. In measuring the continuum of your experience, you become aware of it and become capable of honoring it. In the state of New York portfolios are often associated with so-called alternative schools, ones that use portfolio assessment with students who have struggled in traditional environments. What is happening, obviously, is that in most schools successful students are simply being deprived of a more meaningful (and far less pleasurable) education because of their ability to negotiate slash-and-dash classrooms.

For further research into portfolio assessment, I direct readers to *Portfolio Practices: Thinking Through the Assessment of Children's Work* by Steve Seidel, et al., *Portfolio Portraits*, by Donald H. Graves and Bonnie Sunstein, *Measure for Measure: Using Portfolios in K–8 Mathematics*, by Therese M. Kuhs, and *Constructive Assessment in Mathematics*, by Dave Clarke, all of which are briefly annotated in the bibliography. I also direct them to Brooke Jackson's description of her portfolio process, along with a student sample, in the Samples section.

Sylvia Gross, in the final chapter of this work, expands on the idea that teachers need to be learners as well as students, and enhances our understanding of apprenticeship.

## Works Cited

Berlinski, D. 1995. *A Tour of the Calculus*. New York: Vintage.

Clarke, D. 1997. *Constructive Assessment in Mathematics: Practical Steps for Classroom Teachers*. Berkeley, CA: Key Curriculum Press.

Graves, D. H., and B. Sunstein. 1992. *Portfolio Portraits*. Portsmouth, NH: Heinemann.

Kuhs, T. 1997. *Measure for Measure: Using Portfolios in k–8 Mathematics*. Portsmouth, NH: Heinemann.

Seidel, S., and J. Walters, E. Kinby, N. Olff, K. Powell, L. Scripp, I. S. Veenema. 1997. *Portfolio Practices: Thinking Through the Assessment of Children's Work*. Washington, DC: National Education Association.

# 14

## From Windex to Wildstrom

*Conversations with My Teacher*

### Sylvia Gross

My first year of teaching was a year of silence. Every week night and all day Sunday I would deliver monologues in front of the computer trying to turn my ideas for lessons into feasible plans. Despite helpful colleagues, I would face my class alone and found it easier to keep the plans within the confines of my mind. But teaching is an act of communication, and thrives on conversation. I would eventually find that talking to my students and my colleagues, instead of the sullen screen, would illuminate how to teach. Conversations with my own teacher from high school helped me realize that when I verbalize a problem clearly, my own words shed light on the path to a solution.

That first year I had been teaching fifth grade at Mott Haven Village School in the Bronx, New York. MHVS (or PS 220) is a small preK–8 public school begun by a few elementary school teachers who wanted to create a more intimate, student-centered environment than the one where they had been working. Our students learn from a mix of traditional and progressive teaching methods and approach their studies with enthusiasm and assiduousness. The school's standardized test scores are near the top of our school district, although they trail the rest of the city, state, and country. Continuing to raise the test scores is a major priority for our students, parents, teachers, and school district. In my second year, I transferred into our middle school, where we all teach at least three subjects in order to provide our students with small-group instruction. (This year, my fourth, I taught eighth-grade math, eighth-grade literature, Bronx Studies, and desktop publishing.) Although the breadth of subjects often overwhelms me, I welcome the opportunity to delve into many disciplines.

When I began teaching eighth-grade math I enjoyed it partly because of the seemingly straightforward curriculum, though I'm ashamed to admit it. Teaching English and social studies demanded constant invention. In planning for math, on the other hand, one topic followed another as I flipped through the textbook. More than in the humanities, mathematics demands clear, precise instruction. Carlos, one of my eighth-grade students (all names of students have been changed), announced in literature class that he enjoys poetry because it is mysterious. As I watched him scowling through math class the next day I realized he considered math "mysterious" as well—but hated it. Math instruction is expected to provide access and understanding rather than awe and admiration of mystery. For Carlos' sake, I felt committed to explaining problems clearly. The competent math teacher is basically like Windex, I felt. She makes the solutions to problems so clear that she drops completely out of the frame. Later, the student doesn't remember, "That Mrs. Windex really inspired me," but rather, "I was so smart at math that year!" or "That year math was easy!" For an eager new teacher with a knack for making things more complicated than she had to, following lockstep lesson plans seemed to help.

Yet at the same time, as I struggled to cleanse those minds of doubts and confusion, I had a competing instinct to make the mysteries in math fascinating and enticing with puzzles, relevant situations, and hands-on experiences. These activities occasionally elicited my students' interest but just as often drew out their frustration because I had complicated their learning process. How could I inspire students without also inspiring fear and loathing?

My own math education hadn't prepared me very well to answer this question. The math instruction I had received from elementary school through my years in high school had always been crystal clear but provided very little opportunity for self-direction, with one exception. I enjoyed meeting all the challenges, but the instruction had been too problem based, and I had never become ambitious or creative with the subject. On the other hand, I had always been motivated to read and write on my own, and my humanities education constantly involved independent work. This led me to decide, when I got to college, that literature was my true academic passion. As I pondered how to take my math teaching in a direction it had rarely gone as a student my thoughts turned to the one exceptional math teacher I had learned from: Susan Wildstrom.

Mrs. Wildstrom taught my eleventh-grade class in "honors" elementary functions and analytic geometry at Walt Whitman High School in Bethesda, Maryland. Her difficult tests instilled fear in some, but I remember enjoying the challenge of the assignments and the friendly, methodical way she explained solutions. She always honored multiple approaches to solving problems and provided access to the logic of mathematics through enthusiastic metaphors. (I especially remember developing an appreciation for the multiple ways one could view conic sections: as spatial puzzles, as art, and as symbols.) A classmate of mine was even inspired to develop art projects and poetry based on formulas and graphs from her class.

At the end of my first year of teaching, I wrote Mrs. Wildstrom (as I still felt compelled to call her) a long email, eager to tell her that I had become a teacher and to relate all the details of my struggle. I didn't expect a response necessarily; I just wanted to share my experiences with her. But she wrote back with specific advice on how to improve my instruction. Our conversations began to shed light on the problems that were frustrating me in the classroom.

Our first exchanges were about the strengths and weaknesses of specific students. Linda, for instance, was an immensely literate and otherwise successful student who engaged in a daily battle with mathematics. Though she managed to excel in geometry, where she memorized vocabulary and enjoyed bisecting lines and angles, and she had been able to write an inspired "art criticism" piece based on a postcard of a Kandinsky painting (describing a circle with a 30° slice taken out as "that eminent yellow guy—Pacman!"), she never truly understood the manipulation of numbers. She took outstanding notes—in tricolor—but she consistently used last week's obsolete method to solve this week's problems. To multiply fractions, she found a common denominator. To subtract them, she found the reciprocal. She added time to money with abandon. Although she desperately wanted to achieve the grades in math that she received in every other subject, she refused to engage in the logic of it, as if it were a religion she disdained. She would not look at a problem asking her how many feet it was from New York to Alaska and admit that the answer would have to be a big number.

Susan (as I had now begun to call her) gave me two useful suggestions about Linda. The first was to wait a week to test her on the new material, to give her time to internalize the understanding. The second was to work out the first problem on tests for her, then have her write a paragraph explaining the process—what, specifically, had been done. She could use those directions to continue solving the rest of the problems. As it turned out, Linda was the type of student who needed clearly written steps from the problem to the solution. Using her own writing only made them that much clearer—a perfect use of her writing skills. For other students who had similar trouble performing in test conditions, Susan suggested that I put only one problem per page, or cut up an exam and hand out one problem at a time. This averted the anxiety they felt when confronted by too many tasks at one time.

Eddie was another individual I discussed with Susan. All year long he sat silently where I placed him, front and center of the classroom. His eyes told me, "I will not fail this class although there is no way I could ever get this stuff. I will do everything I am supposed to do but you can't make me think." He occasionally took notes, and his homework was complete every single day, copied off class valedictorian Cristal's work every morning in homeroom. He was proud when he truly conquered one-step algebra problems, but I always became overexcited when he understood something, which made him shut down. Susan suggested that I give him the solution outright (as Cristal already had subversively) and have him use the answers to explain the process in writing.

Giving the students the answer and having them explain the process in writing may have helped Eddie and other students like him, but I also had the likes of James, Anthony, Darren, and Marc in my class. These students were quick learners and loved to reason out word problems together with the whole class. In the evening, though, they tended to lose their enthusiasm and merely scribble some numbers and symbols in their workbooks. It was the communal experience that allowed these students to understand best, one picking up where the other left off as they explained the steps. I sensed a powerful energy in my class when we reasoned as a group like this, but I always feared I was excluding the timid few, who would be thankfully melting in their seats.

When I mentioned my tentative, and somewhat inadvertent, use of this method to Susan, she called it the "chorus approach." To preserve the energy while insisting that the reticent participate, she suggested that I go around the room with each student contributing one step. In another of her strategies, students sit in groups after they have already solved problems independently to "discuss, defend, and argue" the correct answers (Treisman 1992, 362–372). Using this debate technique alongside the independent work would animate my communal problem-solvers as well as engage shy students.

These suggestions built on her idea of putting problems and solutions into words. Over time I realized that many of Susan's ideas involved creating a discussion out of the math problem, using language and logic to understand solutions instead of merely repeating them. From those discussions emerge facility and understanding as each student phrases and rephrases the math in different words. As an added benefit, problem solving becomes a social experience, although each member of a group needs to understand the solution independently for upcoming assessments.

Susan's methods of encouraging dialogue allowed me to make difficult concepts accessible to more of my students. Instead of trying to explain away every problem in front of the class, I could allow them to discuss multiple approaches. This helped make my instruction more effective, but I still wondered how Susan conveyed her passion for the subject and inspired her students to pursue concepts further.

I had often tried to motivate my students by using realistic or appealing situations. The "chorus boys" in my class appreciated the way that I began and ended lessons with word problems (taking surveys of people's television-watching habits, determining how many candy bars Justin eats in a day, measuring how much paper is needed to wrap a box, etc). The concrete examples help make statistics, proportions, or surface area tangible for them, even though the problems may not be truly relevant to their lives. I asked Susan for her thoughts about the issue of relevance in math. She was wary of too much reliance on relevance, per se, and was critical of poorly written textbooks that, simply to engage the students, "are filled with pictures, historical notes and 'application' problems the basis for which, in many cases, the kids don't have." She herself loves math for "its beauty and logic. The fact that doing a problem several different ways should (and does) give the same answer."

In her emphasis on thorough understanding, she inspires students by helping them conquer their fear, which in turn empowers them. She described a two-year algebra course, called Algebra Part I and Algebra Part II, that gave the students two years to grasp the concepts. "I wish we could bring this course sequence back. I enjoyed teaching it because the slower pace allowed kids to learn the material. By the end of the second year, they thought the teacher was a 'god' because she had finally been able to teach them math. They didn't always realize that the pace had finally given them the time to really absorb the concepts." Mrs. Windex was not that far off—a little clarity goes a long way to motivating students.

Not long after reading Susan's letter about the issue of relevance, I went to a meeting of teachers who were discussing effective assessments for students across subjects. One teacher, in explaining that assessments should be relevant to the students' lives, ended her commentary with, "After all, who ever uses calculus in everyday life?" Before better judgment could set in I shot back, "People who know how to do calculus." Since when are engineers, research scientists, architects, economists, and math teachers not a part of everyday society? Clearly, the teacher in question does not expect her students to be able to rise to those professions. She ignores the fact that no one will be able to apply, let alone enjoy, calculus if they have never mastered it.

Often, rather than motivating students by struggling to make activities relevant to lives, we might help them achieve mastery with completely "irrelevant" games and puzzles. (What can be more fun than utter nonsense?) One of Susan's own students pointed out after reading Martin Gardner's "A Quarter Century of Recreational Math" (1998) that by "learning with a game . . . knowledge can be gained at a much faster rate. When students want to learn and are not dreading the teacher's next lecture, the environment is much more conducive to gaining knowledge." Although the perfect formal game to teach a specific concept or skill is sometimes hard to come by, students also derive satisfaction ("Oh, snap!") from simple surprises, such as the discovery that algebra will automatically give you an answer that you had to make twenty guesses to solve otherwise, or the sudden recognition of the pattern in a sequence of numbers that had looked random before. The satisfaction these puzzles inspire is not infantile. When we solve them we are often looking directly at the "beauty and logic" of math itself.

Susan also uses a kind of game to teach students to be creative and investigative in their thinking: "My husband travels to a lot of technology shows and conferences. He gets a lot of 'stuff'—t-shirts, stopwatches, pens, clocks, gizmos, etc.—that he brings home and gives to me. I take them to school. Whenever a student makes an interesting comment or observation, something exceptionally perceptive or a way to solve a problem that I haven't seen before but that is especially effective, I say 'a t-shirt or a prize for . . . ' and at the end of class, they can select an award for their insight. No one ever knows in advance what sort of remark is likely to precipitate an award. Most of my students are anxious to offer observations and guesses in class discussions. Last year one of my kids said 'This place is like a carnival!'"

I have discovered on my own that the prize itself isn't important. My eighth graders will jump up and down for a lollipop, sticker, or a "check plus" in my grade book. What is important is that they learn what a "smart" comment in math really is. They get a clearer picture of what they should strive for in their own work. The value is placed on astute observations, creativity, and insight, not merely the "right answer."

My early discussions with Susan thus involved helping struggling students achieve mastery through explanation and conversation; however, my class last year also included the likes of Gabriel, who needed only to see the first step of a problem to intuitively understand the rest of the lesson for the day and race through his class work and homework with spare precision. (Gifted math students seem to enjoy solving problems using the minimum number of steps.) When I described Gabriel to Susan, she responded, "What a delight! Did he often get frustrated with how long it was taking the rest of the class to catch on? Was this a problem?" Of course he did occasionally get annoyed, but the students in my school generally don't disrupt when they understand sooner than the others do. As a result, the brightest students may not receive the attention they deserve. "He is the kid for whom you probably want to have math books to read for fun," Susan advised.

Susan is the first to recognize that students who ace regular assignments need an added challenge. Although most of her students at Whitman are eager to do well, some stop trying when they get the right answer, especially in math. Susan aptly described this tendency of some Whitmanites to be, "grade grubbing and no smarter than they need to be." By contrast, the middle school teachers at Mott Haven try to instill the "grade grubber" in our students in order to keep them from losing interest entirely; however, the intellectual apathy Susan encounters ranges from "grade grubbing" to systematic and remorseless cheating on tests.

Many teachers might consider this form of mathematical malaise inevitable, but Susan has found a way to gently prod her students into feeling a personal and intellectual investment in their study of math. Language arts teachers typically encourage their students to expand their interests by reading, writing, or seeing plays and films beyond the curriculum. Independent essays based on personal selections of literature are common. If an English class focuses on memoirs, a teacher may ask each student to read one of their own choice and write an essay on it. Students develop a personal relationship with the field and a sense that their work is unique. When I had her as a teacher, Susan was no less prepared to provide ways for us to challenge ourselves through book suggestions, competitions, and extra courses. Imagine a typical math teacher saying, "You liked that? Now find a geometry book that interests you and solve some more problems"—and yet Susan would do precisely that. (See the annotated bibliography on the Internet at *http://www.heinemann.com/shared/ onlineresources/E00245/bibliography.pdf* for Susan Wildstrom's recommended reading and website lists.)

In fact, this type of reading (and writing) assignment has now become an even greater component of her curriculum. As Susan explained in a talk delivered at the Mathematical Association of America in 1999, she created reading assignments in response to her school's call for increased "literacy across the curriculum (Wildstrom 1999,1)." Coincidentally she had herself become frustrated that all her students wanted out of math was a right answer. Wanting them to see it as more than "numbers and problems and answers," she devised a reading assignment that calls for her students to read one article, chapter, or book about mathematics (at least 10 pages) of their own selection each quarter. They now regularly tackle a range of literary, scientific, historical, and recreational approaches to mathematics with enthusiasm. Popular titles are William Dunham's *Journey through Genius, Time Travel and Other Mathematical Bewilderments* (and other books by Martin Gardner), Raymond Smullyan's *The Lady and the Tiger,* Eli Maor's *e: The Story of a Number,* and books by Ivars Peterson and Ian Stewart. Susan is amused by the fact that so many students consider it an "easy A" or apologize for reading things that seem "simple." Their dynamic responses show how she met her goal of "wanting kids to learn for themselves that it is possible to read mathematics for pleasure, the same way they read fiction and history for enjoyment."

Many of the readings initially appear intimidating and dense to her students, but a little concentration usually leads to clarity and interest. One wrote, "I am fascinated by the evolution of the human mind, and while I wouldn't always apply that interest to math, this assignment has given me a rare opportunity." Another was excited at finding a discussion of a Chinese puzzle game she had played in Chinese school. Others read about the implications of game theory on political science or marveled at the mathematical patterns in nature. Students pursued music, time travel, literature, and art—all using the tools of math. Some practiced their math skills in puzzle books, whereas those with more confidence in their abilities tackled more sophisticated proofs and problems. Reading their comments made me want to look for these books and go learn more math.

Susan's recent assignment sheets in reading and writing follow. For her recommended lists of books and websites, see the annotated bibliography on the Internet at *http://www.heinemann.com/shared/onlineresources/E00245/ bibliography.pdf*

### Reading and Writing in Mathematics

Each quarter there will be a reading assignment. You will read something mathematical that is both interesting and comprehensible to you. The choice of what you read will be mostly up to you. I will establish some basic guidelines and if you want to deviate significantly from them, I would ask you to speak with me personally about your intent.

The assignment will be as follows: select a mathematical "article" of approximately ten or more pages in length. Read it carefully, more than once if necessary so you understand what it is presenting. Then write about a page (the equivalent of a double-spaced typewritten page—250 words) in which you state clearly what the source of your reading is and explain the article, making it clear to me that you have understood the material.

You will be graded as follows: This assignment will be worth 25 points in each quarter. You will generally receive full credit if (1) you fulfill the above criteria, (2) you convince me in your written work that you have read and understood the material you have chosen, and (3) you submit your work on time. The assignments will be due on the first school day that is on or following (1) November 1, (2) January 1, (3) April 1, and (4) June 1 (seniors should submit this on or before the last time they attend class with me). This work will be accepted early if you wish.

The material you read should be of an appropriate challenge based on the math course you are taking. However, it should also be on a subject that interests you, even if that is not one your course involves. The "ten-page" criteria is simply to prevent students from selecting one-page "quickies" such as are found in *The Joy of Mathematics* or other books of this sort. If you wish to dip into books that have a lot of small articles on a variety of subjects, feel free to do so, but select about a dozen of them. On the other hand, should you decide to "read" (study) a very difficult proof of a sophisticated hypothesis and it only runs a page in length, there will probably be no reason why I wouldn't encourage you to use that (but please be sure that you consult me about this first).

In addition to the library, other fine sources of such materials can be found in the bookshelves of our classroom and in our school library and if you are looking for a specific book or a specific subject, I can check to see if the department owns a book that would suit you. There are lots of books on subjects outside the objectives of our courses—number theory, n-dimensional geometry, Fibonacci numbers, logic puzzles, problem-solving techniques from the masters, history of mathematical development, etc. I also have a rather nice collection of journals that I have received over the years. Some of the articles in them would make interesting manageable reading as well. Browse through the collections. I will also publish a list of some selected book titles that I think you might find enjoyable.

Articles may be found on the Internet, however they are often not of appropriate length, so you are more likely to find good materials for this assignment in printed matter.

### Using Information Technology

Another required assignment each quarter will involve the use of the computer in locating mathematical resources. Each quarter, you will be required to submit evidence that you have "visited" one or more "web sites" related to mathematics. Again, the choice of site will be largely yours, however the material contained at the site you choose must relate in some direct and discernible way to a topic we are studying in your course during the quarter for which you use it. There are a number of sites that provide math challenges and problems that you can submit online answers to; there are sites that will do your homework for you (especially in calculus). There are sites that will tell you more than you ever wanted to know about mathematical subjects—prime numbers, pi, etc. I will try to maintain some sort of list of the URLs (annotated as much as possible) for web sites that you might visit, especially those that will provide you with good links to a variety of other sites. If you find a site that is particularly special, I would hope that you would add its URL to our ongoing list.

As with the reading assignment, you will receive full credit provided you give evidence that you have visited a site, give a complete URL for the site, used the resource for its intended purpose, and gained something by the experience and you submit this on time. Due dates will be as with the reading, the beginning of the month near the end of the appropriate quarter. A printout of one screen from the web site and your brief comments about the site including how it relates to your course of study will be an appropriate submission for this assignment.

### Scoring Deductions for Reading and Internet

2.5 points will be deducted from each assignment for each day that it is turned in late (standard 10% deduction policy)

#### *Reading*

If the selection is too short (extremely difficult material may be fewer than ten pages if discussed with teacher), 2 points will be deducted per page short of the suggested 10 pages.

If the written submission is too short (the expectation is approximately 250 words) 3 points will be deducted.

If the written submission does not convince me that you understood what you read, up to 5 points will be deducted.

#### *Internet*

If the URL is not written out clearly and completely (I should be able to use your URL as written to go to the same site you went to), 3 points will be deducted.

If the printout of a screen is missing, 6 points will be deducted.

If your own comments on the site are not written, 6 points will be deducted.

Another way Susan encourages the students' personal investment in their work is through math journals that they send to her via email. The journals allow Susan to keep tabs on her students' relationship with the course and also serve as an assessment of her own instructional activities. It is important to note that all these assignments supplement an already challenging math curriculum. I recall from Susan's class that test questions came almost exclusively from the most difficult problems we encountered in class and on homework assignments (a practice I emulate when I create my own tests). As a result, these "ancillary" assignments must come as a godsend to some of Susan's intimidated students.

Although the assignments may seem like less of a challenge, they also reveal hidden avenues of access to the mathematical content of the course.

Here is Susan's email journal assignment:

### Email Journals

During each quarter, please send me (not all of them at once but spaced out to roughly every two weeks) five Email journals in which you tell me how you see your progress in this course going. They can be as long or short as you feel is necessary for each one. But they should be specific in describing what is working or not working for you about the class, what sorts of activities are helpful, fun (?), etc. Each journal will be worth 10 points (for just doing it), with the total for each quarter's journal assignment being 50 points. (You may communicate with me by Email as often as you wish, however, when your Email represents what you see as a journal entry—and you are not limited to five—please use the word "journal" in the subject line).

As a suggestion for a very useful "first" journal, I would enjoy reading a mathematical autobiography in which you would tell me a bit about your mathematical background—any interesting experiences or discoveries that you have made, any activities in which you have participated that have somehow shaped your attitudes toward math, what your attitudes toward math are, any individuals (with or without names) who have somehow influenced your attitudes (positively or negatively) toward mathematics. You control the tone and length.

Both the readings and the journals provide opportunities for Susan to have more in-depth and wide-ranging mathematical "conversations" with her students—increasing not only their comprehension but also their intellectual engagement. Not only are they communicating their ideas with her, they are entering into a mathematical discourse that is normally the domain of college and graduate students. It is a logical extension of her approach of "putting a problem into words."

Although I initially felt uncomfortable with spending time on math journals in my own class, I began using them this year in various ways. Because my students are younger than Susan's, I feel they need more directed assignments and deadlines. At first, I had students explain why certain "tricks" worked (why you can simply move the decimal point in a number in order to convert metric units, for example). I found that most students simply explained the procedure instead of why it worked. It did, however, spotlight the students who had an unusually strong comprehension of the mathematical basis for the steps. I was most impressed when I had students review a test in their journals and explain why they got certain questions wrong. They threw themselves into their explanations and the correct answers as if they were defending themselves to a judge!

The danger with literacy assignments comes when they replace skill building and mathematical content. Teachers uncomfortable with math can use

reading or writing as an excuse to engage in what President Bush might refer to as "fuzzy math." A colleague of mine jokes that sixth graders use art
and writing in their math journals to explain basic arithmetic concepts they
mastered in second grade. One of my first attempts to use writing in math was
the "art criticism" piece I mentioned earlier. Although it seemed like a good
idea, I was frustrated by the fact that ninety percent of the students did not incorporate the grade-level vocabulary I wanted them to use (similar, congruent,
transversal, corresponding, . . . ) but rather used very elementary terms (line,
circle, square, . . . ). This year I transformed the assignment so that the students
drew their own geometric art pieces and, like the painter Sol Le Witt, wrote
explicit directions on how their pieces were created. They then exchanged directions and attempted to re-create their classmates' designs. Many of the redrawn
pieces turned out exactly the same as the originals. On the other hand, some
students were frustrated by the lack of precision in the language and, in disgust,
returned the directions to the writers for clarification. In the resulting conversation, the students realized that a consistent vocabulary was key to mathematical
communication and that in order to put mathematics to work (building bridges,
designing graphics, or teaching it!) they needed to have a firm grasp of terms
as well as concepts.

I have mentioned already that Susan has been finding her students to be
less prepared in recent times. It has been useful for me to see her response.
Rather than disparage them, she has invented procedures to ensure their intellectual honesty and, even more importantly, has developed a curriculum in
which students are able to pursue their own mathematical interests independently (just discovering that they have their own mathematical interests is a
great deal!). In her talk about the reading assignment, Susan described one
student: "What I like most about her is her honesty and determination. Mathematics doesn't come particularly easily to her, but by effort she has been very
successful. Her readings and her reaction to them seem to show that she ap-
preciates the powerful ideas that underlie what we all take for granted." As a
result of these assignments, Susan has become more impressed with their personalities and abilities. In an environment where students are being pressured
to get high grades, she shifts the emphasis to reward curiosity and intellectual
engagement.

What she has learned, and passed along to me, has had an equal impact
on my impressions of my students. Sharing their idiosyncrasies, strengths, and
weaknesses with Susan helped me realize that I could tailor my instruction
to the individuals and engage them in conversations that pave the road from
problem to solution in words.

How did she become a teacher who so naturally seems to know how to
explain and inspire? As early as eighth grade, Susan remembers teachers who
inspired her to want to teach math. She was identified as a gifted math student in a
high school that was reorganizing its program. Therefore, she occasionally spent
semesters studying math with only a few other advanced students on her grade

level. Caught in between courses, they taught themselves a curriculum at their own pace. Thus she had an early experience with self-directed instruction. She also took enrichment courses over several summers through a program of the National Science Foundation. Her natural curiosity and passion for math were fired through her opportunity to study independently and encounter challenging material.

In terms of her own mentors, Susan described a professor she had in college whose "down-to-earth" style she emulates in her own classroom. She mentions specifically his "theoretical remarks interspersed with good examples that almost looked as though they were being made up on the spot." (I remember admiring the very same spontaneity in her class.) Susan recalled, "He would sometimes pose questions in the context of his lecture and casually say, 'Take that as an exercise.' It wasn't until the third or fourth week, after some problem sets had been collected, that he realized that we had interpreted the remark to mean that we should merely think about the question. He really meant for us to solve those problems as part of our homework. So he started putting problem numbers on these propositions and we dutifully copied them down so we would have them later. I just sensed a complete command of the material, an easygoing attitude about the classroom setting that encouraged us to feel comfortable and to be willing to ask questions and offer ideas." Her recollections mirror the classroom she conducts today, particularly her aside that the informality of his class applied to the atmosphere rather than the rigor of instruction.

Susan also continues to be involved in mathematics and math education. She collects math books ("some of which I buy for their beauty, classic treatment, or historical value") and actively participates in many professional organizations. In this way she contributes to the national conversation on the direction of math education. Whether the topic is a student, an instructional method, a political issue, or a math problem, Susan applies the same blend of humor and acuity that make her an intelligence to be reckoned with and at the same time, approachable.

Although I didn't use all of Susan's techniques immediately, her ideas and comments stayed with me throughout my day in the classroom and during my weekend planning. Eventually, not only her specific ideas but her approach began to affect my own practice. All her innovations (journals, readings, insight incentives, and group problem-solving) encourage high-level discussion requiring that mathematics be used logically and comprehensibly.

Over the past few years, I have opened myself up to conversations with other colleagues. I spent many hours a week bothering my very patient principal and exchanging ideas with Tammy Vu, a friend who began teaching at the same time (who also contributed an article to this book). My teaching grew through these discussions in which I chronicled my reactions to the classroom, discussed my difficulties, compared my experiences with others, and sorted through many solutions to improve my own teaching. This dialogue itself is what pushes us (both the students and me) toward mastery.

The Jewish philosopher Abraham Heschel proclaims that education is not only for us to understand but also to "stand still and behold! Behold not only to fit what we see into our notions, behold in order to stand face to face with the beauty and grandeur of the world" (Heschel 1966, 42). If we want students to devote time and energy to pursuing ideas instead of just points, grades, and dollars, we need to find ways for students to independently investigate and discuss what intrigues them. They almost always find something interesting when they begin to look, listen, and speak.

## Works Cited

Gardner, M. 1998. "A Quarter-Century of Recreational Mathematics." *Scientific American* 279(2): 68–76.

Heschel, A. 1966. "Children and Youth." *The Insecurity of Freedom: Essays on Human Existence.* Philadelphia: Jewish Publication Society of America.

Treisman, U. 1992. "Studying Students Studying Calculus: A Look at the Lives of Minority Math Students in College." *College Mathematical Journal.* 23(5): 362–372.

Wildstrom, S. 1999. "Encouraging an Enjoyment of Mathematics through Reading and Writing." Mathematical Association of America.

# Conclusion

Sylvia's story serves as *Teaching for Depth's* denouement. Writing, reading, speaking, listening, contextualizing, historicizing, investigating, modeling, reiterating, revisiting, and revising—all are featured in this one tale of a teacher's coming to learn that to teach she must learn and that to learn she must come out of her shell and converse with others. The thing about this engagement is that it happens one gear at a time: One teacher noticing one student's needs. One student noticing one other student's academic behavior. One teacher talking to one other teacher. "Each one teach one" was an adage of the sixties. So it is here.

Sylvia underscores the stakes: How much suffering results from lack of engagement, in loneliness, frustration, and fear. How much potential for pleasure there is when we do get engaged. One of her students "refused to engage in the logic of math, as if it were a religion she disdained." Another hated the arbitrary mysteries of it. How did Sylvia cope with these devastating responses in her classroom except to seek help, to "open myself to conversations"? How did her students overcome their responses except to do the same, whether in written or spoken form?

William Carlos Williams defines a poem as "a small (or large) machine made of words. When I say there is nothing sentimental about a poem I mean that there can be no part, as in any other machine, that is redundant. . . . Its movement is intrinsic, undulant, a physical more than literary character. . . ." (1969, xv). We are saying, similarly, that learning is a machine made of conversations, in a literal and symbolic sense. There is equally no sentiment about it. The participation, in all its modes, can and should be measured until it is as unsentimentally efficient as possible. Does this steal the pleasure of teaching? On the contrary. How many of us have been forced to sit in classrooms with no point to the work, no focus to the conversation, no correlation between what we learned and the grades we were given? Every minute spent in such classrooms was wasted, and insufferable, especially considering how fulfilled we could have been if teachers and administrators had taken the time to measure, and thereby to see, the true nature of their students' experience.

It is on this note, of the students' experience, that we come to the end of our journey across the disciplines. The journey being over, we have time to reflect on its scope and meaning. For me it actually began when I was a student in the 1960s at Robert E. Lee High School in Baton Rouge, Louisiana. I remember guidance counselors advising me to capitalize on my work as editor of

*The Traveller*, the student newspaper, and focus on a major in English when I got to college, the better to prepare myself for work in public relations. Though I had done well in math and enjoyed the sensation of computing answers (for that was pretty much what our math studies were all about in those preconstructivist times) I was told it would be a different beast in college, far too difficult, and I should avoid it. Thirty years later, having dodged that career in public relations to become a writer and writing teacher, I found myself back in the high school classroom, now as a staff developer.

Memories of the initial fulfillment of editing *The Traveller* guided me in my work as much as anything. The schools I worked in had adopted, in their constructivist pedagogy, the very principles of teamwork and relevance embodied by that formerly "extracurricular activity." Often students were as engaged in their learning as I had been in the breaking news. Helping them achieve a sensation of pleasure in their learning became as gratifying as anything I had ever done. I began to notice, though, how difficult it was to incorporate mathematics into our interdisciplinary projects. Math teachers often wanted to get involved but didn't have the time to break away from their regimented curricula. The students I talked to, like so many in this book, had become conditioned to think likewise—that math was separate and didn't belong anywhere but the math classroom.

My high school guidance counselors' augury that I would make an excellent PR man came to mind. Though they hadn't had much imagination about career choices, they had done well to steer me away from the cruelty of relentless drilling in mathematics. The trouble was they had inadvertently steered me away from math stories, from math history, from the greater perspective on life one has from being able to see through a mathematical lens. It occurred to me that my present work offered an opportunity to recover some of these lost pleasures for myself and in the process perhaps to deliver them to the students. I imagined the students in history class discussing the debt democracies owed to Kepler, Galileo, Copernicus, who broke down the walls of orthodoxy with their math-based science. I imagined them not coming out of their calculus classes saying "Math is so *boring!*" (a comment I overheard in the halls recently) but in a state of visceral wonder at the way a thrown ball moves more slowly as it approaches the clouds, descends more rapidly as it reaches the grass. They marveled, in my imagination, that the ball's speeds during the changing moments can actually be measured, that its arc as it travels its parabolic path can be predicted, that the area under the path can be calculated, and that the individuals who discovered this were two of the greatest minds of the Renaissance. If those students were inclined toward careers in public relations, they might consider that the corporations who hired them probably wouldn't exist without the political as well as scientific changes wrought by Newton and Leibniz.

Having had good experiences a decade before when I was commissioned to coauthor *The Art of Science Writing* (Worsley and Mayer 1989), I began to talk to people like Ginny Cerussi and Matthew Szneher about a book on mathematics. As the conversation extended to the other contributors a clear thesis began to emerge. It might be phrased thus: *The teaching of math can make sense to students and its pleasures can be recovered if its isolation, its "otherness," is understood and solved.* It was this thesis that guided our inquiry and that led to the inclusion of our four areas of discussion. The thesis also demanded an investigation into the history of mathematics' "otherness."

It is not new. In the ancient world, Egyptian priests kept their astronomical calculations secret to protect their power. The Hellenistic Greeks were loath to see the abstract beauty of their mathematical discoveries sullied by real-world application. Renaissance philosophers Hume and Berkeley struggled to keep mathematics separate as a method of interpreting the material world even as Hobbes and Locke were advocating the supremacy of math in its power to elucidate reality. In the seventeenth century Descartes complained about the deadening effect of math's isolation in these words:

> When I first applied my mind to Mathematics I read straight away most of what is usually given by the mathematical writers, and I paid special attention to Arithmetic and Geometry because they were said to be the simplest and so to speak the way to all the rest. But in neither case did I then meet with authors who fully satisfied me. I did indeed learn in their works many propositions about numbers which I found on calculation to be true. As to figures, they in a sense exhibited to my eyes a great number of truths and drew conclusions from certain consequences. But they did not seem to make it sufficiently plain to the mind itself why these things are so, and how they discovered them. Consequently I was not surprised that many people, even of talent and scholarship, should, after glancing at these sciences, have either given them up as being empty and childish or, taking them to be very difficult and intricate, been deterred at the very outset from learning them.... But when I afterwards bethought myself how it could be that the earliest pioneers of Philosophy in bygone ages refused to admit to the study of wisdom any one who was not versed in Mathematics ... I was confirmed in my suspicion that they had knowledge of a species of Mathematics very different from that which passes current in our time. (Descartes 1912, 2)

In the early twentieth century, from the other side of the curricular equation, playwright George Bernard Shaw complained about instruction he received at the Wesleyan Connexional School:

> Not a word was said to us about the meaning or utility of mathematics: we were simply asked to explain how an equilateral triangle could be constructed by the intersection of two circles, and to do sums in $a$, $b$, and $x$, instead of in pence and shillings, leaving me so ignorant that I concluded that $a$ and $b$ must mean eggs and cheese and $x$ nothing, with the result that I rejected algebra as nonsense, and never changed that opinion until in my advanced twenties Graham Wallas and Karl Pearson convinced me that instead of being taught mathematics I had been made a fool of. (Pearson 1942, 48)

We saw a similar articulation in the mid-twentieth century in Morris Kline's comments, quoted from my chapter with Kay Rothman, "A Mathematical Correspondence Between Humanists." He goes on to say, as we observed in the preface, that "[s]ince the layman makes very little use of technical mathematics, he has objected to the naked and dry material usually presented. Consequently, a subject that is basic, vital and elevating is neglected and even scorned by otherwise highly educated people. Indeed, ignorance of mathematics has attained the status of social grace" (Kline 1953, vii).

Our thesis demanded us to understand the origins of mathematics' "otherness," and it also required us to understand those features of its curricular isolation that are legitimate. John Berry demonstrated in his Lab workshop the value of pure mathematical investigations, which generate the mathematical thinking that leads to better modeling. In a casual conversation about the subject a colleague used the metaphor of a banquet. "Soups and salads do belong in the same meal," he said, "but you don't put the lettuce in the soup." You do, however, swing your tennis racket later using calories from all the courses. In school, we rarely investigate reality at such a level. This is the level we contend must be explored.

We realize that it takes a significant change in perspective to make the exploration possible. I liken it to my own experience confronting non-Euclidean geometry in the course of my research for this book. As I read passages on the subject in *Mathematics in Western Culture*, particularly those debunking the axiom on parallels (which says that through a point P not on a line L there passes one and only one line M—in the plane of P and L—that does not meet L no matter how far M and L are extended) I had a visceral reaction of anxiety. In fact, in the margin next to the sentence "These reflections suggested to Riemann that he adopt along with the finiteness of the straight line an axiom to the effect that there are no parallel lines" (423). I wrote, "Still saying get outta here. (Feeling sick.)" But, as I wrote in the margin a couple of pages later, "My whoa is changing to wow!" Euclidian geometry makes a lot of sense to a carpenter, but it is not the ultimate definition of geometric reality. Similarly, math classes make a lot of sense for developing students' ability to think, but they do not define all there is to know about mathematical reality.

I remember when I confronted Einstein's general theory of relativity for the first time and discovered that though Newton's gravitational laws, based on Aristotelian concepts of time and space, accurately predicted the behavior of matter, they were based on false assumptions. The sun doesn't reach out with some invisible force (gravity) through some invisible substance (the ether) to hold the planets in their orbits. They are all located on one continuum of space–time. What the contributors to this book are proposing is that, as educators, we become aware of the continuum that connects the subjects just as the space–time continuum connects the sun to the planets. Yes, study the behavior of the heavenly bodies discretely when it makes sense to do so, but remember that the division is ultimately artificial.

We are obviously not the first to have the idea of an educational continuum. John Dewey underscored the importance of students' "experiential continuum" (Dewey 1997, 33). Piaget, Vygotsky, and their successors have effectively sustained and elaborated the idea, and it is beginning to emerge in many progressive and reformist programs. The movement for performance standards, the International Baccalaureate's Primary and Middle Years Programmes, the Coalition for Essential Schools, Expeditionary Learning Outward Bound, the Education Alliance—these are but a few. All, at one time or another, have either confronted or will have to confront the beastly side of math's isolation.

The rewards for a successful confrontation will be great. We've only begun to discover them in this work. I wish it could continue. For instance, I am now reading Eli Maor's *To Infinity and Beyond* and have just been contemplating the concept of the Continuum Hypothesis (66), how it shook the foundation of mathematics in becoming an axiom that one is free to accept or reject. Doesn't such a choice defy the very idea of an axiom? I wonder if there is any possible correlation between the continuum of infinities and the continuum of experience. I wonder if the story of Georg Cantor might not be dramatized—for the way the scalding criticism of his ideas on the hierarchy of infinities contributed to the depression that drove him into a mental institution at the end of his life. A kind of mathematical Van Gogh, he was appreciated only later by luminaries like Bertrand Russell, who said, "The solution of the difficulties which formerly surrounded the mathematical infinite is probably the greatest achievement of which our age has to boast." David Hilbert said, "No one shall drive us from the paradise Cantor created for us (Maor 1987, 64)." Haven't we room, then, to explore this amazing personage? What, too, about the stirring political and psychological dramas in the story of Évariste Galois, the tormented French republican who met his death by duel at the age of twenty, or the intriguing protofeminist career of the physicist and mathematician Sophie Germain?

There seem an infinite number of ideas and observations that bear mentioning in support of our thesis and the programs that understand it. The thesis itself could probably evolve infinitely. Fortunately, however, like a limit in calculus, theses mandate an arbitrary terminus in narrative compositions. Sanity lives here in this end zone if we can curb our momentum long enough to reflect on what we may have accomplished instead of what we might continue to do. Looking at our thesis as a kind of hypothesis, or conjecture in need of proof, we might see if we have achieved our proof. Let us contend that by pleasurably experiencing math's natural relationship to the humanities, students will (1) understand the nature of reality more deeply, (2) develop an abiding sense of curiosity in the world around and within themselves, and (3) feel confident that they can research from multiple points of view the questions that come to them as they mature. Let us posit further that we will have succeeded if in response to our setting out on a journey across the disciplines, a significant number of students become emboldened to embark on similar journeys of their own.

Have a significant number of students done so? In Ginny Cerussi's experiments "Fifteen students responded positively, three responses were negative,
and one student was unsure if the journal writing had helped her." In my experiments with Matthew Szenher, all the students professed to enjoy the writing,
and half found it of some use mathematically. In John Berry and Susan Picker's
research, "Where 40 (24.7 percent) of their prepanel drawings had depicted
stereotypical images of weirdos, wizards, Einstein-like figures, or had been
missing an image, this style of depiction was reduced to just 13 (7.4 percent)
of drawings. . . . [W]here originally 37 percent of the girls had drawn a female
mathematician, 51.1 percent now depicted a female." In many cases we didn't
measure but relied on anecdotal responses, which revealed that at least some
students were positively affected by our efforts.

Luckily we are not alone in the field. The more I researched it, the more
activity I found. One particularly steady beat comes from Alvin White and
the Humanistic Mathematics Network. The masthead on the journal mentions
most of the central figures working directly in the humanistic mathematics
movement. A search of the Internet reveals dozens of sites where good work
is being done. Recently, for instance, I came across an essay by David Dennis
at the University of Texas at El Paso, entitled "The Role of Historical Studies in Mathematics and Science Educational Research" (2000). He sets out
three clear directions for historical and educational research in mathematics,
around the notions of context, content, and critique. Stephen I. Brown's work
in the humanistic mathematics education movement brings considerable bril-
liance to the common effort. Matthew Szenher discusses Brown extensively
in the annotated bibliography (on the Internet at *http://www.heinemann.com/
shared/onlineresources/E00245/bibliography.pdf*) In fact, all the authors in the
annotated bibliography and the "Samples" section also support our cause, either
directly or indirectly.

To me the stakes in our combined efforts are very high. In today's *New
York Times* (2/10/02), in a front page article on upcoming challenges to the
constitutionality of the voucher system in Cleveland schools, I read the following statement: "barely a third of public school students graduate from high
school." The voucher system in effect advocates combining church and state
as a solution to this problem. We advocate combining the disciplines in meaningful ways to reflect reality. Maybe this, along with other features of school
reform, will change the statistics. I compare the alienation of mathematics from
the humanities to the kind of prejudice we encounter too often in the political
and social arenas. Rage builds up around fixed ideas because the continuum
of reality is no longer sensed or understood. Rigidly compartmentalizing the
academic disciplines parallels that process because it numbs us as learners and
teaches us that walls between ways of thinking are natural when, in actuality,
the structure of our mind is fluid. This does not mean students shouldn't delve
deeply into the disciplines. On the contrary, it suggests that if they are allowed to
delve deeply, they will find the hidden, but essential, connections. Why should
students have to rebel to avail themselves of the wholeness of this experience?

In the spring of 2001 I was pressed into service monitoring New York State's standardized eighth-grade math test. I saw the symbol $\pi$ at the top of a page. Its number value was given to the students to help them solve problems on the test. The symbol looked rather bare and naked above the busy text below. Students quickly referred to it as they hurried through to produce the answers the test demanded. As I watched them labor I reflected on my own experience with pi—how in reading books about it, I had marveled at its seeming infinitude of amazing properties, how it had led me to rich conversations about literature and science with other teachers, how I wondered, as had Lawrence Cinamon in Matthew Szenher's chapter on apprenticeship, if it could be the base of a number system. At the end of their allotted time, the students handed in their test booklets to the proctoring teachers and escaped to fast-food restaurants in the neighborhood. I signed my papers warranting that test conditions had been met, left the room, and forgot about what I had observed until the following October, when I read in the Metro section of the *New York Times* the following headline: "Majority of Eighth-graders Again Fail Statewide Tests." I couldn't help but ask, "Would so many have failed the test if it had guided teachers to teach the properties and history of the number as well as its mere utility in solving equations? What if they had been prepared to respond intelligently to Nobel Prize–winning poet Wislawa Symborska's poem 'Pi'?" In the poem she talks about pi's infinite number of digits that travel, in an unremittingly random pattern, "over a wall, a leaf, a bird's nest, clouds, straight to the sky/through all of the bottomless bloated heavens . . . (2001, 32)."

In *Teaching for Depth* we advocate getting to the tested math skills through a sense of wonder in the subject. We invite readers to imagine a continuum of educational experiences where literacy is universal, where math may be found wherever it is relevant, where students are given permission to ask the questions that arise naturally and to answer them with support, not resistance, from the disciplines. We urge teachers to overcome their anxieties and follow this vision on a journey across the hall so their students can find their own roads to *wisdom*, as Descartes would have it.

# Works Cited

Dennis, D. 2000. "The Role of Historical Studies in Mathematics and Science Educational Research." In *Research Design in Mathematics and Science Education.* Mahwah, NJ: Lawrence Erlbaum.

Descartes, R. 1912. *Discourse on Method.* New York: E. P. Dutton. Cited in Morris Kline, *Mathematics in Western Culture* (New York: Oxford University Press, 1953).

Dewey, J. [1938] 1997. *Experience and Education.* Reprint, New York: Touchstone.

Kline, M. 1953. *Mathematics in Western Culture.* New York: Oxford University Press.

Maor, E. 1987. *To Infinity and Beyond: A Cultural History of the Infinite.* Princeton, NJ: Princeton University Press.

Pearson, H. 1942. *G.B.S. A Full Length Portrait*. New York: Harper.

Szymborska, W. 2001. "Pi." In *Numbers and Faces: A Collection of Poems with Mathematical Imagery*, ed. J. Growney. Claremont, CA: Humanistic Mathematics Network, p. 32.

Williams, W. C. 1969. *William Carlos Williams: Selected Poems*. New York: New Directions.

Worsley, D., and B. Mayer. 1989. *The Art of Science Writing*. New York: Teachers & Writers.

# Samples

At the beginning of this project I envisioned a book on writing in the math classroom. The mission grew into its present larger form, but writing remains the main delight I experience in relation to mathematics. The language of mathematical operations is by necessity condensed to the purest and least ambiguous expression possible; but the language of mathematical thinking and feeling is a site of great ambiguity, detail, and nuance. Already in this volume we have encountered many forms of reflection and expression in mathematics. The following samples of math writing extend the repertoire of ideas teachers might consider for use in their work with mathematical ideas. If it were 1000 pages long, it would not be comprehensive. Like the annotated bibliography, the samples should be seen as models for the products that might emerge in the kinds of humanities and mathematics classrooms we have discussed in the main text. They constitute only a small, condensed part of what I have been able to assemble in my "journey across the curriculum." Again, I merely hope that the section will help teachers to begin, the better to give their students a start.

## Amy Quan Barry

Amy Quan Barry, writing in the second person, accentuates the incalculable equations of our emotions by describing mathematically those things around "the poet" that seem, under the circumstances, to be more calculable. Students might be asked to use the language and symbols from math class to describe landscapes, and they might discuss in their writing what can, and what can't, be calculated in both their inner and outer worlds. As in all the samples, students might use the form of the model in their writing. In this case, Barry is writing in three-line stanzas (tercets) of free verse.

IF $\frac{dY}{dX} = \frac{4x^3+x^2-12}{\sqrt{2x^2-9}}$, THEN

you are standing at the ocean,
in the moon's empirical light
each mercurial wave

like a parabola shifting on its axis,
the sea's dunes differentiated & graphed.
If this, then that. The poet

laughs. She wants to lie
in her own equation, the point slope
like a women whispering stay me

with flagons. What is it to know the absolute value
of negative grace, to calculate
how the heart becomes the empty set

unintersectable, the first & the last?
But enough.
You are standing on the shore,

the parameters like wooden stakes.
Let $x$ be the moon like a notary.
Let $y$ be all things left unsaid.

Let the constant be the gold earth
Waiting to envelop what remains,
the sieves of the lungs like two cones.

# Samuel Beckett

Beckett writes with mathematical precision in general and sometimes includes mathematical ideas in his texts. A scatological scene in *Molloy* looks at math with ironical humor even as it makes observations about homeless life and takes a satiric shot at a journal of the day:

> And in winter, under my greatcoat, I wrapped myself in swathes of newspaper, and did not shed them until the earth awoke, for good, in April. The Times Literary Supplement was admirably adapted to this purpose, of neverfailing toughness and impermeability. Even farts made no impression on it. I can't help it, gas escapes from my fundament on the least pretext, it's hard not to mention it now and then, however great my distaste. One day I counted them. Three hundred and fifteen farts in nineteen hours, or an average of over sixteen farts an hour. After all it's not excessive. Four farts every fifteen minutes. It's nothing. Not even one fart every four minutes. It's unbelievable. Damn it, I hardly fart at all, I should never have mentioned it. Extraordinary how mathematics helps you to know yourself.

In his *Collected Shorter Prose 1945–1980* we find a somber short work called "All Strange Away," in which the narrator describes his tomblike dwelling. The dimensions begin at "Five foot square, six high, no way in, none out ... " and grow smaller by carefully calculated degrees, each time with references to the coordinates of the corners. Another character's space is described thus:

> Cease here from face a space to note how place no longer cube but rotunda three foot diameter eighteen inches high supporting a dome semi-circular in section as in the Pantheon at Rome or certain beehive tombs and consequently three foot from ground to vortex that is at its highest point no lower than before

with loss of floor space in the neighborhood of two square feet or six square inches per lost angle and consequences for recumbent readily imaginable and of cubic an even higher figure, all right, resume face. But a, b, c and d now where any pair of right-angled diameters meet circumference meaning tighter fit for Emma with loss if folded as before of nearly one foot from crown to arse and of more than one from arse to knees and of nearly one from knees to feet though she still might be mathematically speaking more than seven feet long and merely a question of refolding in such a way that if head on left cheek at new a and feet at new c then arse no longer at new d but somewhere between it and new c and knees no longer at new b but somewhere between it and new a with segments angled more acutely that is head almost touching knees and feet almost touching arse, all that most clear.

That Beckett is at first somewhat difficult and often scatological does not mean his work can't be broken into useful chunks as writing models. Imagine students inventing characters who experience physically the mathematics the students are learning. In this case the geometry of the shrinking tomb is correlated with the parts of the body in a way that is ultimately as musically poetic in style as it is precise in measurement.

# Rita Dove

Rita Dove celebrates her success at her mastery of mathematics in this poem. Students might pause once a week to celebrate their learning and their accomplishments, with the proviso they write with the quality and style modeled here.

### Geometry

I prove a theorem and the house expands:
the windows jerk free to hover near the ceiling,
the ceiling floats away with a sigh.

As the walls clear themselves of everything
but transparency, the scent of carnations
leaves with them. I am out in the open

and above the windows have hinged into butterflies,
sunlight glinting where they've intersected.
They are going to some point true and unproven.

# Elizabeth Fox

Elizabeth Fox, proponent of nonlinear literary practices, as evidenced in her contribution to this volume, "Encouraging Chaos," wrote the following prose poem. It plays on associations with the theoretical concept in physics of absolute zero, or complete absence of heat, and the concept in mathematics of absolute value. Students might similarly let the associative meaning of mathematical terms guide them in writing a poem about their sensations of reality.

### Absolute Zero

I live alone. Let the value of my apartment be that of an ocean liner where the city is the Atlantic during a storm. The city moves around me. A thunderclap sets off car alarms outside. I hear a metal sign from a gas station tumble down the street.

Inside my living room I stand on pointe and spin. The absolute value of any part of me exists in relation to a point of balance. When I find this point my body is fixed to the earth. If the line that describes the number zero is a membrane, let it expand to include my entire body. I can stay here as long as the music allows.

What music? The ghost of John Cage takes charge of the weather. Wind and gas station signs, car alarms and thunder accept him as their conductor. He commands a pause. For a moment the night, the city, the ocean, this apartment, are silent. The agitation of matter ceases, and its molecules come to rest. Call this moment absolute zero. Let all sound before and after take value in relation to it.

The thunderstorm outside continues. Wind loosens a metal sign from its moorings and pushes it down the street. When the wind lifts it up, the sign is silent. I hear it land again and again.

I stop spinning and come off pointe. In the absence of absolute zero, information from the senses becomes tricky. For what seems like days I am lost at sea. As I look out the window of my apartment I see only the horizon and clouds moving across the sky abnormally fast. This is not my first voyage.

I become aware of the treacherous distance between my wooden floor and the ground, and imagine that if I try to leave my room I will find an abyss where my doormat used to be.

The air is full of moving particles. In some places they crowd together. Even so, they keep moving. The denser areas signify an edge. I've seen particles pass through my door. Those that belong to the wood change places with those that belong to the air. Over the years I've come to understand that it is in my best interest to act as if these edges were solid.

# Galileo Galilei

Too often we study subjects in school without hearing the voices of the thinkers who have conceived them to begin with. To hear their voices is to feel the life being breathed back into material that may have become dry and academic. The following passage by Galileo (translated by Henry Crew in 1914) is stunning in its simplicity and clarity as he sets out to "set forth a very new science." Indeed, the Renaissance may be said to have reached full force with the phrase "I have discovered by experiment. . . ." Certainly, Isaac Newton felt this when he spoke of standing on the shoulders of giants to, as Galileo put it, "explore the remote corners" of the new science. (See John Maynard Keynes' biographical material on Newton in this section as well.) Students might use features from this passage when introducing research papers or lectures to fellow students: the statement of purpose, the setting of historical context, and the element of personal reflection.

### Third Day
### Change of Position [De Motu Locali]

My purpose is to set forth a very new science dealing with a very ancient subject. There is, in nature, perhaps nothing older than motion, concerning which the books written by philosophers are neither few nor small; nevertheless I have discovered by experiment some properties of it which are worth knowing and which have not hitherto been either observed or demonstrated. Some superficial observations have been made, as, for instance, that the free motion [naturalem motem] of a heavy falling body is continuously accelerated; but to just what extent this acceleration occurs has not been announced; for so far as I know, no one has yet pointed out that the distances traversed, during equal intervals of time, by a body falling from rest, stand to one another in the same ratio as the odd numbers beginning with unity.

It has been observed that missiles and projectiles describe a curved path of some sort; however no one has pointed out the fact that this path is a parabola. But this and other facts, not few in number or less worth knowing, I have succeeded in proving; and what I consider more important, there have been opened up to this vast and most excellent science, of which my work is merely the beginning, ways and means by which other minds more acute than mine will explore its remote corners.

## Brooke Jackson

Earlier we discussed portfolio practice as a kind of calculus by which students measure their own engagement. Brooke Jackson, an English teacher at the New York City Lab School for Collaborative Studies, structures her portfolios with the following "equation," articulated in a curricular letter to parents:

*Evaluation and Assessment:* Students will engage in a *reflective portfolio process* in which they will: Set *individual goals* and track their process of approaching them, and consider and work towards embodying our *shared guiding principles*: risk-taking, conventional excellence, process, and growth. Our portfolio process culminates at the end of each quarter when students compose a *reflective cover letter* that traces their experience and performance. Cover letters, from a student's perspective, tell the complex story that underlies the numerical grade that they propose, having carefully accounted for—relying on their *contracts*: an individual log similar to a column in a traditional teacher's grade book—their work in relationship to individual goals and shared guiding principles. Following completion of the quarter's portfolio, a one-on-one *conference* allows student and teacher to review the portfolio together, consider and potentially negotiate the proposed grade, reflect on the quarter past, and begin to negotiate the next quarter's goals.

Here are a tenth-grade student's two sample cover letters, from successive quarters:

### REFLECTIVE COVER LETTER—Q2
### Allegra Warsager—201

I had tighter expectations for myself this quarter, and I tried to express that in my Q-2 goals. I wanted to make sure that the faults I had last quarter (resulting in me getting an 80 as my grade) wouldn't be faults again for the 2nd quarter. However, I don't think I did the amazing job I had hoped to do. My main area of problems is/was turning in work late. I'm approaching that by getting an organizational tutor—hopefully it will work. One of my other goals was to work more productively with my response group, and I did a good job meeting that goal. I offered comments, suggestions, and paid attention to what everyone had to say. When we shared our Metamorphosis essay drafts, I tried to read as many as I could and responded in a way that I hoped would help them. As for any other goals, I never felt like my actual writing needed improving, just getting it in. . . .

In my writing I took risks by basically not caring what people thought of my writing—like if it was inappropriate, or fake-ish, or boring. I wrote whatever I wanted and continued to ignore any negative comments that would bring me down when I tried to write. For conventional excellence, I don't mean to sound "proud" or like I'm bragging, but that was never really something I had a problem with. Process and growth are constant things in writing, and with each piece I wrote, no matter if it was a scrap or an essay or a progress report, I could feel new ideas clamoring to get out of my brain—ideas that had never been there before and were brought on by what I had learned writing about previous things. My process was sometimes none at all—I wrote things I didn't want to change, not even one sentence. But sometimes my process was heavy—there were things that I wanted to say that had potential, but were expressed badly and needed to be added to/fixed.

One problem that I've had that has been ongoing for the whole semester is the vocab quizzes. I don't mind studying for them, but it's difficult to remember all the definitions. I think I need a better way of studying for them. Something that I've grown to like much more over this semester is scrap assignments.

Each topic gets more challenging, but that's what I need, otherwise writing becomes a bit of a joke for me if the topic is too easy. Things like attendance and participation have never been much of a problem for me, although I am not a perfect student in that sense. Sometimes I won't say anything during a class, because there is not much for me to say. I hope that next quarter I'll be able to find more things to participate with, but I'll save that for my Q-3 goals. As for major writing pieces, there was only one for this quarter, and although I handed it in late (as usual—this must be stopped) I feel like I finally did a five paragraph essay that I didn't dread doing, and that I was proud of. Now I'm prepared for others, which I know will come soon.

### REFLECTIVE COVER LETTER—Q3
#### Allegra Warsager—201

My main goal this quarter was to improve my organizational skills: handing in work on time, making sure I did everything that needs to be done. And of course I didn't do an absolutely perfect job, but I did *a lot a lot a lot a lot* better than in the 2nd quarter and I'm proud of myself that I could do that and how much harder I worked to do that.

I took a lot of risks this quarter in my writing like I do every quarter, but I think I went above and beyond my old writing this time. And I got a lot more comfortable with my class so I wanted to share more with them. I worked on process more also—I edited everything on a piece that I thought needed editing, and I used any advice I could get that I thought would be helpful. That also helped me with growth as well because with all the changes I made, I started to become more critical of my grammar and writing, etc. The books I've read so far this year have helped me grow a lot too, and also the one-word scraps: "Underground," "Truth," etc. I never really do well on the vocabulary tests because I have a bad memory for things like that, but I'm going to try harder for the last quarter. My response group this semester was soooo much better than it used to be I feel like my response group actually takes an interest in my writing and what I have to say more than my old response group did, and we all get along better in general. As for participation/attendance, I get sick a lot, but I always make sure I make up what I miss, and I participate any time I have something I want to say—which is like all of the time, especially during text talks.

I would give myself around an 88 this quarter I guess … well, anywhere from an 85–88. I don't exactly know how well I did. All I know is that I've improved a lot and my grade probably (hopefully) went up the grade I just proposed. I LOVE ENGLISH!!!!!

After her grading conference with Brooke, Allegra received a 90 for the quarter. Students who have overestimated themselves might agree to a lower grade, knowing exactly how and why they earned the number ascribed to their work. See the bridge to Sylvia Gross' chapter "From Windex to Wildstrom: Conversations with My Teacher" for more thoughts on portfolio assessment.

# Rodger Kamenetz

If you were a number, which would you be? What properties would you have?
What number would a friend, a historical figure, an author be, and why? These
are some of the potential prompts suggested by Rodger Kamenetz' poem
"Mallarmé in Tournon." Note that Kamenetz takes the point of view of the
historical figure about whom he writes—the French symbolist poet Stéphane
Mallarmé. Note also the setting of the school and the attitude of the narrator
toward his job as schoolmaster.

### Mallarmé in Tournon

Nothing
Just the blue
Not a dot on the horizon
Dinner's done and the last noises
of the village have died down
I am not even tired
Today they didn't starve me with questions
or wring the last ounce of feeling from my skin
Monsieur P—— even says I show "promise"
as a schoolmaster

I have not yet shown them my queer lazy side
I am reasonably attentive to their announcements

Because I am young they expect more to come
They don't know I'm burning at night
the fabulous coal of their future
They have plans
for they arrange the desks in rows
The pupils who tread the maze
will have first chance to rearrange it
Happy permutations!
No guess who will wind up in front
Perhaps some day I will stumble into place
and success will pay me back
Until then I make only
the most obvious calculations
As I am outside the sum
I can be added on indefinitely
without effect, like a zero

Zero had to be invented
The other numbers were merely practical
apples and oxen
but zero was the truth of their predicament
the glittering varnish of their sum
And I am zero, hors-concours
Though I don't abide mistakes in my salary

# Claire Leavitt

High school student Claire Leavitt wrote the following movie review for her math club newsletter at the New York City Lab School for Collaborative Studies. It is a good model of student use of the form. Movies and plays with mathematical subjects and themes abound—from Broadway's *Proof* and *Copenhagen* to Off-Broadway's *Hypatia* to films ranging from *Pi* to *A Beautiful Mind*. Mathematics is often romanticized as an inaccessible practice of geniuses or crackpots in these films, providing fodder for student debate in light of Susan H. Picker and John S. Berry's work on the image of mathematicians. Another idea is to use the idea embedded in the movie itself: to write a story, play, or screenplay in which math is used as a device.

### *The Cube*

Six people wake up one morning and discover they are trapped in a large cube. The cube is divided into hundreds of cube-shaped rooms with a door on each wall. Every door is numbered and they lead to rooms filled with deadly traps. Some rooms are simply death traps. There is only one escape route, and the possibility of freedom is lessened by the fact that every room in the cube is continuously shifting and being rearranged like a gigantic Rubik's cube.

The premise of *The Cube* is intriguing and the plot description was what attracted me to the film. While having mathematical content, the movie also includes a complex character study, pieces of action, violence, and a portrait of desperation and longing for freedom.

There are six main characters in the film. At the beginning of the movie, no one can tell whether or not all of the characters will make it out of the cube alive. Yet the question of which characters will survive continuously changes and keeps the viewer on the edge of their seat. As the film progresses, the audience comes to see the true faces of each of the characters. Their masks are slowly torn away by fear, anger, and desperation.

*The Cube* uses mathematics to involve the audience and keeps it riveted. For example, Leaven a teenage girl who is among six people trapped inside the cube, is an excellent math student. Therefore, the others rely on her intelligence and abilities to help them all escape. Leaven concludes that every trap-free room has a prime number marked on its door, which makes it easy for her to figure out which rooms are safe to enter. Other areas of mathematics, most specifically geometry, are also important to the plot.

The level of violence in the film is quite shocking, ranging from the actions of certain characters to the merciless traps in the cube. However at the end of the film, I realized that the violence was necessary in order to really understand the movie. As a whole the movie is thrilling, gripping and challenging to the mind. Some scenes are very heavy with dialogue, which gives a nice contrast to the very violent scenes.

Weaknesses in the film include the poor lighting, which in a movie like this one can make it all the more confusing. The camera work is decent considering that the entire film takes place inside the cube. The highlight of *The Cube* is the way it makes the audience think. The audience is never told who built the cube, where it is, why these six people were put there, or where they will be if they are ever able to escape. Even the obvious desire of the characters to escape the cube begs the question: Escape to what? The unexpected conclusion even further emphasizes this philosophical conundrum. The film is best not seen with your brain on idle. *The Cube* is very unique film and I think it is safe to say you've never seen anything like it.

# Diane Lefer

Math functions to define character in this complex scene from Diane Lefer's story "Mr. Norton's Wart Hog." What is your, or your character's, relationship to math? might be a guiding question for student writing. Note here that for Ethan math makes the world a place of peace. It is a way to file down the rough edges of reality. For Julie, it is a way for the girl to get closer to Ethan. Another idea is to use this passage as a model for students to describe their own experience with math homework.

### Excerpt from "Mr. Norton's Wart Hog"

Julie Maldonado was not a genius, as Ethan soon found out when he returned on Tuesday evening after work to tutor her in algebra. They cleared space at the bar—a little bit of jungle claimed for civilization and homework. Julie had a maddening tendency to guess at answers and no intuitive feel for numbers. Neither did Ethan; he thought his own life could inspire her. Bounced from town to town as a kid, school to school, no one had expected much of him, but somehow he'd found that numbers balance, that there could be order and harmony to the world. Math homework had given him a peaceful feeling and accounting had saved him. He'd embraced a conventional life and he had everything he wanted—a C.P.A. license, a suburban home, a wife he loved, and Claire was talking now about a family. (She was also very tolerant of his adventures. According to Claire, Ethan had almost no imagination—a shame, as a person is the better for a rich inner life. If Ethan could only expand his horizons, then, by tagging along behind strange people and keeping secrets, that was all right with her. Apparently it was the best he could do.)

"Be painstaking," he told Julie. "The world we live in is imperfect, there's little we can do to make it right. But here, on this piece of paper. . ."

Julie had no intention of following his advice. Still, she liked Ethan, or rather she liked men. On retreat with a group from school, a priest had talked to them and explained how at their age, even nice boys are after only one thing; they can't help it. Ethan, being grown up, was presumably after something else. Though Julie could not imagine what it was, she trusted his age.

"I didn't even have your advantages," he said. "Nature didn't bless me with good looks and charm—or even a great deal of ambition. But I learned to work hard with what I've got."

"X equals 24y?" she guessed, twisting her pencil in her hair. Ethan groaned, but that didn't bother her. He didn't seem at all ready to give up.

# Raye Levine

Raye Levine was an eighth grader at the Clinton School for Writers and Artists when she wrote and graphed the following fairy tale in her math class. Her teacher, Augustina DiGiovanna, had been inspired, in part, by Kurt Vonnegut's essay later in the section. Writing such stories has the potential to deeply ingrain math skills and concepts, particularly as students approach the abstractions of algebra and the idea of functions.

### The Hummingbird and the Crane

A long time ago, when animals were claiming their territory on the earth, the hummingbird and the crane were having a dispute. They were discussing who would get to drink the fresh sweet water of the pond nearby. They needed to figure out a way so someone would have it, because they could not share. Then the hummingbird suggested a race between the two of them. The hummingbird said that was his final offer and whoever won the race would get the pond for good. The crane thought about it for a while and then agreed. The race was about to begin the next day. The hummingbird was known for his fast flying. He could go about 200 miles per hour. The crane on the other hand didn't have it so good, because he could only fly close to 25 miles per hour. And despite the hummingbird's arrogance, the crane was confident and determined to win over his pond. They set up a spot to meet the following day, and negotiated all of the rules and regulations of the race.

The next day came within the blink of an eye, and sure enough when the crane got to their meeting spot, the hummingbird was there to greet him. The race began, and already the hummingbird was out of sight. The crane held out his large wings while keeping a strong and steady pace. Then the crane saw beautiful flowers up ahead and when he got closer to passing them, he got a whiff of that sweet-smelling nectar, but he was determined to win the race and he didn't let anything distract him or get in his way. The hummingbird, on the other hand, knew that the crane didn't stand a chance and stopped to drink some of the nectar from the sweet-smelling flowers that he passed by on the way to the finish line. Soon enough the hummingbird grew very tired from the juice in the flowers and stayed for what seemed to be a short moment to take a rest. Then he started up again, and after he went a little ways towards finishing, he saw nobody behind him and figured the crane was still far behind, so he started feeling hungry again, and he drank all of the nectar he could. He then felt tired again, and fell fast asleep. The crane came close and passed by the hummingbird. The crane kept going and shortly after he passed the hummingbird, he got to the end of the race. When the hummingbird woke up, he flew right over to the finishing line to find the crane there waiting, exhausted, out of breath. The hummingbird couldn't believe that he had actually lost the race, and congratulated the crane for winning his own pond. Then the crane said that the hummingbird could choose what he wanted besides the pond and the hummingbird chose the nectar from the sweet-smelling flowers he had passed. And that's how the hummingbird owns the nectar of the flowers (see Figure).

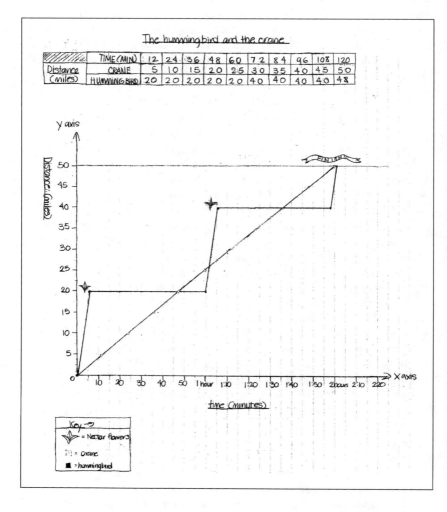

The race between the hummingbird and the crane was 50 miles long. It takes the crane 2 hours to fly the whole race without stopping (since he goes at a speed of 25 miles per hour) and it takes the hummingbird 15 minutes to fly the whole race without stopping (since the hummingbird goes at a speed of 200 miles per hour). So to make this graph, I had to make a range, of the most amount of time spent between both the hummingbird and the crane and the most distance made. What that information, I set up my graph. I then made a chart that showed the crane's distance after $x$ amount of time and the hummingbird's distance after $x$ amount of time. After that, I graphed my data. In the beginning, I had to make ratios to find out the distance for the amount of time gained, in order to put down the chart and make my graph. The hummingbird could have easily won the race in 15 minutes, but he was so sure he was going to win, he knew he had time. Unfortunately, he fell asleep twice from the sweet nectar juice of the flowers. The first time he rested for 54 minutes, then the second time he fell asleep for 52 minutes, which was too long, and lost the race by 2 minutes. At the end of the story the crane gets to keep the pond, and the hummingbird gets the nectar flowers. So, although the hummingbird loses the race, all ends well.

# John Maynard Keynes

In 1933 the great economist John Maynard Keynes produced a work called *Essays in Biography*, in which he treated the life of Isaac Newton. Whereas Bertrand Russell's memoir is excerpted later to provide a model for how students might write about themselves, Keynes' essay gives us a model for how students might write biographical sketches of one another, perhaps establishing a thesis then providing anecdotal support, complete with dialogue, as Keynes does. Note that the essay inadvertently touches on one of the important underlying themes of *Teaching for Depth*: the education of the intuition. It also vouches for the power of inquiry and experimentation in learning.

I believe that the clue to his mind is to be found in his unusual powers of continuous concentrated introspection. A case can be made out, as it also can with Descartes, for regarding him as an accomplished experimentalist. Nothing can be more charming than the tales of his mechanical contrivances when he was a boy. There are his telescopes and his optical experiments. These were essential accomplishments, part of his unequalled all-round technique, not, I am sure, his *peculiar* gift, especially amongst his contemporaries. His peculiar gift was the power of holding continuously in his mind a purely mental problem until he had seen straight through it. I fancy his pre-eminence is due to his muscles of intuition being the strongest and most enduring with which a man has ever been gifted. Anyone who has ever attempted pure scientific or philosophical thought knows how one can hold a problem momentarily in one's mind and apply all one's powers of concentration to piercing through it, and how it will dissolve and escape and you find that what you are surveying is a blank. I believe that Newton could hold a problem in his mind for hours and days and weeks until it surrendered to him its secret. Then being a supreme mathematical technician he could dress it up, how you will, for purposes of exposition, but it was his intuition which was pre-eminently extraordinary....

There is the story of how he informed Halley of one of his most fundamental discoveries of planetary motion. "Yes," replied Halley, "but how did you know that? Have you proved it?" Newton was taken aback—"Why, I've known it for years," he replied. "If you give me a few days, I'll certainly find you a proof of it"—as in due course he did.

# Jody Madell

Ninth-grade global history teacher Jody Madell has been working with "accountable talk" scoring forms and rubrics for several years at the New York City Museum School. This is the latest evolution of one of her more comprehensive forms.

**Record of class discussion:**

    Gives information from reading: "R"
    Information from another source: "S"
    Makes comparison: "C"
    Questions: "Q"
    Idea/Synthesis: "I"
    Responds to another student: "A"
    Distraction/Disruption: "D"

| Student | Date: 9/11 | Date: 9/17 | Date: 9/20 | Date: 9/24 |
|---------|-----------|-----------|-----------|-----------|
| S— | R | | | |
| E— | | A | | R |
| F— | I | D | | RRR |
| A— | IRI | ISS | S | RCQ D |
| E— | | | | |
| C— | ARC | | | |
| A— | D | R | | RRAQ |
| F— | | S | | |
| A— | IRC | RII | | RR |
| M— | I | S | | |

| | | | | |
|---|---|---|---|---|
| N— | RR | R | S | D |
| G— | | D | | |
| J— | DRRRC | RCS | | A |
| T— | | | | |
| N— | Q | | S | II |
| C— | R | RI | | R |
| M— | SR | AR | | Q |
| L— | I | | | |
| L— | | A | S | |
| J— | | R | | |
| M— | C | A | | R |
| A— | | | | R |
| I— | | | | Q |
| Z— | | | | |
| A— | | R | | RR |
| B— | S | SS | | IIR |
| G— | R | | | |

# Howard Nemerov

As does Jonathan Swift in a later passage, Nemerov sees geometry in the world around him. He refers to the Platonist's argument that mathematics is a "pure Euclidian kingdom of number". The ironic use of a biblical tone focuses our thoughts on our isolation from the primal forces in our lives. The poem might well be used as a point of entry to a debate about the nature of mathematics (perhaps on Dewdney's guiding question, "Is mathematics created or discovered?"). Students might also simply write from the questions, Where do you see geometry in your life? What does it mean?

### Grace to Be Said at the Supermarket

That God of ours, the Great Geometer,
Does something for us here, where He hath put
(if you want to put it that way) things in shape,

Compressing the little lambs in orderly cubes,
Making the roast a decent cylinder,
Fairing the tin ellipsoid of a ham,
Getting the luncheon meat anonymous
In squares and oblongs with the edges bevelled
Or rounded (streamlined, maybe, for greater speed).

Praise Him, He hath conferred aesthetic distance
Upon our appetites, and on the bloody
Mess of our birthright, our unseemly need,
Imposed significant form. Through Him the brutes
Enter the pure Euclidian kingdom of number,
Free of their bulging and blood-swollen lives
They come to us holy, in cellophane
Transparencies, in the mystical body,

That we may look unflinchingly on death
As the greatest good, like a philosopher should.

# Plutarch

One of the central messages of *Teaching for Depth* is that history belongs in the mathematics classroom, and vice versa. Students are often encouraged to write biographies of mathematicians as extra-credit projects. We would have such projects integrated more centrally into the curriculum, and we would provide the most literate models of the form. Such models serve more purposes than merely to acquaint us with the facts of historic lives. They provide depth of understanding, of the mathematics as well as of human nature. In this excerpt from Plutarch's biography of Marcellus, a Roman military commander who has since become largely forgotten, we get insights into Archimedes, the emergence of Platonism, and the very schism we are trying to bridge in this book. We see Archimedes not only as a great mathematician and inventor but as a humanist who wanted to bring the pleasures of mathematics to "people in general."

### From "Marcellus"

Marcellus, with his sixty galleys ... furnished with all sorts of arms and missiles ... assaulted the walls [of Syracuse], relying on the abundance and magnificence of his preparations, and on his own previous glory; all which, however, were, it would seem, but trifles for Archimedes and his machines. These machines he had designed and contrived, not as matters of any importance, but as mere amusements in geometry; in compliance with King Hiero's desire and request, some little time before, that he should reduce to practice some part of his admirable speculations in science, and by accommodating the theoretic truth to sensation and ordinary use, bring it more within the appreciation of people in general. Eudoxus and Archytas had been the first originators of this far-famed and highly prized art of mechanics, which they employed as an elegant illustration of geometric truths, and as a means of sustaining experimentally, to the satisfaction of the senses, conclusions too intricate for proof by words and diagrams. . . . But what with Plato's indignation at it, and his invectives against it as the mere corruption and annihilation of the one good of geometry,—which was thus shamefully turning its back upon the unembodied objects of pure intelligence to recur to sensation, and to ask help (not to be obtained without base subservience and depravation) from matter; so it was that mechanics came to be separated from geometry, and, repudiated and neglected by philosophers, took its place as a military art. Archimedes, who was a kinsman and friend of King Hiero, wrote to him that with a given force it was possible to move any given weight; and emboldened, as it is said, by the strength of the proof, he averred that, if there were another world and he could go to it, he would move this one.

# Bertrand Russell

Enlightened classrooms encourage students to see themselves as practitioners of the disciplines they are studying. They often write biographical sketches of themselves as historians, scientists, poets, and the like. Here is a passage from Bertrand Russell that proves an excellent model for such biographies. It is an excerpt from his 1944 memoir *The Philosophy of Bertrand Russell*. (See the earlier excerpt by John Maynard Keynes on how students might similarly write biographical sketches of one another.)

> A great event in my life, at the age of eleven, was the beginning of Euclid, which was still the accepted textbook of geometry. When I had got over my disappointment in finding that he began with axioms, which had to be accepted without proof, I found great delight in him. Throughout the rest of my boyhood, mathematics absorbed a very large part of my interest. This interest was complex: partly mere pleasure in discovering that I possessed a certain kind of skill, partly delight in the power of deductive reasoning, partly the restfulness of mathematical certainty; but more than any of these (while I was still a boy) the belief that nature operates according to mathematical laws, and that human actions, like planetary motions, could be calculated if we had sufficient skill. By the time I was fifteen, I had arrived at a theory very similar to that of the Cartesians. The movements of living bodies, I felt convinced, were wholly regulated by the laws of dynamics; therefore free will must be an illusion. But, since I accepted consciousness as an indubitable datum, I could not accept materialism, though I felt a certain hankering after it on account of its intellectual simplicity and its rejection of "nonsense." I still believed in God, because the First-Cause argument seemed irrefutable.

Note how Russell entwines his philosophical predilections and his feelings into his observations about mathematics. He is talking about mathematics' effect on *himself*, giving us a glimpse into the continuum of his experience, as Dewey would have it.

# David Schubert

In his poem "The Mark," the talented, short-lived, psychologically conflicted poet of the 1930s David Schubert describes his sense of himself in terms of grades. It is a harsh critique of the letter system of grading as well as an inadvertent alphabet poem (in which a letter of the alphabet becomes the subject of a poem). Students might enjoy discussing their own evaluations in a poetic light, making them more aware of the math of their own engagement.

**The Mark**

Sad as the rain am I now that God has
Graded me with a $B-$; in his class,
I loved the recess, studied the window.
Is it my fault who built me that way? Yet
Even God must suffer at his mistakes.

Why did he lie? Or didn't he know
Who promised me, that of phoenixes, I
Was not to be cremated, but a Glory.

A $B-$ hurts; it isn't even
A mediocrity; not an $A$ standing there
On its own legs, a smart man; but
A curved Greek, pliant and polite,
Lacking something.

Think of the sinuous bosom
Of a $C$, which sees all, and feigns
Indifference! An open mind is a $C$, a good
American, friendly, someone you can talk with. A $D$
On the other hand, stands for
Damn you! Who

Can survive its scurrilous echo?
And $E$ is like an eel, squashy, squishy—
But mud in your eye whichever way you look at it.

As for the sacred excommunication's
$F$—final is it, finalities
Beyond the grave. And like the question *why*,
Haunting the victim in his tabula rasa.

# Cheryl Shafer

A seventh-grade math teacher at the New York City Lab School for Collaborative Studies, Cheryl Shafer uses various reflective practices in her classroom to ensure the "continuum of experience" of her students. The following simple tool helps prevent the Zeigarnick effect (discussed earlier by Kay Rothman in her chapter "A Mathematical Correspondence Between Humanists.") after students take tests. (See Virginia Cerussi's chapter "Laying the Foundation: Writing in the Math Classroom" for further examples of reflective writing in the math classroom.)

| Date | Question | Original Answer | Correct Answer | Explanation of Error | Who Helped? |
|---|---|---|---|---|---|
| 11/1/01 | 6.b. If a family has 3 children, what is the probability that all 3 children are boys? | 3/6—girls 3/6—boys | The probability that all 3 children are girls is 1/8. The probability that all 3 are boys is 1/8. | Since my sample space list was incorrect, that means that this answer would be incorrect—caused by miscounting. | Figured out on my own |
| | 6.c. What is the probability of having 2 girls and 1 boy? | 3/6 | The probability is 3/8. | Answered above | Figured out on my own |

## Jonathan Swift

As did Howard Nemerov earlier, Jonathan Swift, in "Voyage to Laputa" from *Gulliver's Travels*, used geometric figures to describe food:

> My Dinner was brought, and four Persons of Quality, whom I remembered to have seen very near the King's Person, did me the Honour to dine with me. We had two Courses, of three Dishes each. In the first Course, there was a Shoulder of Mutton, cut into an Æquilateral Triangle; a Piece of Beef into Rhomboides; and a Pudding into a Cycloid. The second Course was two Ducks, trussed up into the Form of Fiddles; Sausages and Puddings resembling Flutes and Haut-boys, and a Breast of Veal in the Shape of a Harp. The Servants cut our Bread into Cones, Cylinders, Parallelograms, and several other Mathematical Figures. While we were at Dinner, I made bold to ask the Names of several Things in their Language; and those noble Persons, by the Assistance of their Flappers, delighted to give me Answers, hoping to raise my Admiration of their great Abilities, if I could be brought to converse with them. I was soon able to call for Bread, and Drink, or whatever else I wanted.

Scholars have questioned why Swift lampooned mathematics so severely in his great satire. During his time mathematics was coming out of its ivory tower to enlighten understanding of nature and to bring all forms of cultural expression down to earth, so to speak, from excessive floweriness and irrelevance. Whatever his reasons, Swift did skewer Platonic ideas of mathematics with his rapier wit. Students might use satire and mathematical imagery to comment on their own experience of artificiality in their environment. Another of my favorite passages from Swift comments on educational practices that are still too commonly encountered in schools today:

> I was at the Mathematical School, where the Master taught his Pupils after a Method scarce imaginable to us in *Europe*. The Proposition and Demonstration were fairly written on a thin Wafer, with Ink composed of a Cephalic Tincture. This the Student was to swallow upon a fasting Stomach, and for three Days following eat nothing but Bread and Water. As the Wafer digested, the Tincture mounted to his Brain, bearing the Proposition along with it. But the Success hath not hitherto been answerable, partly by Error in the *Quantum* or Composition, and partly by the Perverseness of Lads; to whom this Bolus is so nauseous, that they generally steal aside, and discharge it upwards before it can operate; neither have they been yet persuaded to use so long an Abstinence as the Prescription requires.

Swift's texts are obviously wonderful documents for the study of history and the development of sophisticated reading skills, as well. His shocking and hilarious satiric essay "A Modest Proposal," also driven by math (statistics, in particular) gets a fine reaction from high school students who at first don't know it is written as satire.

# Kurt Vonnegut

This essay by Kurt Vonnegut provides a hilarious model for graphing one's life, movies, plays . . . stories of any kind. (See earlier the work of eighth-grader Raye Levine for a student sample inspired, in part, by Vonnegut's work.)

### Excerpt from *Palm Sunday*

What has been my prettiest contribution to my culture? I would say it was a master's thesis in anthropology, which was rejected by the University of Chicago a long time ago. It was rejected because it was so simple and looked like too much fun. One must not be too playful.

The thesis has vanished, but I carry an abstract in my head, which I will here set down. The fundamental idea is that stories have shapes which can be drawn on graph paper, and that the shape of a given society's stories is at least as interesting as the shape of pots or spearheads.

In the thesis, I collected popular stories from fantastically various societies, not excluding the one which used to read *Collier's* and *The Saturday Evening Post*. I graphed each one.

Anyone can graph a simple story if he or she will crucify it, so to speak, on the intersecting axes I here depict:

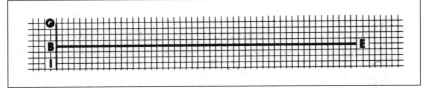

"G" stands for good fortune. "I" stands for ill fortune. "B" stands for the beginning of a story. "E" stands for its end.

The late Nelson Rockefeller, for example, would be very close to the top of the G–I scale on his wedding day. A shopping-bag lady waking up on a doorstep this morning would be somewhere nearer the middle, but not at the bottom, since the day is balmy and clear.

A much beloved story in our society is about a person who is leading a bearable life, who experiences misfortune, who overcomes misfortune, and who is happier afterward for having demonstrated resourcefulness and strength. As a graph, that story looks like this:

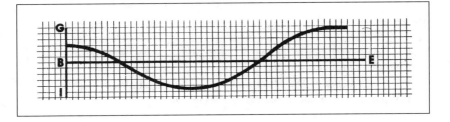

Another story of which Americans never seem to tire is about a person who becomes happier upon finding something he or she likes a lot. The person loses whatever it is, and then gets it back forever. As a graph, it looks like this:

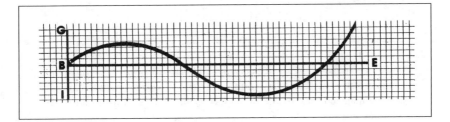

An American Indian creation myth, in which a god of some sorts gives the people the sun then the moon and then the bow and arrow and then the corn and so on, is essentially a staircase, a tale of accumulation:

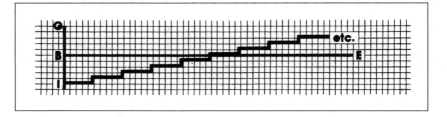

Almost all creation myths are staircases like that. Our own creation myth, taken from the Old Testament, is unique so far as I could discover, in looking like this:

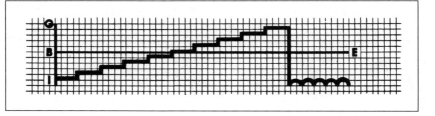

The sudden drop in fortune, of course, is the ejection of Adam and Eve from the Garden of Eden.

Franz Kafka's "The Metamorphosis," in which an already hopelessly unhappy man turns into a cockroach, looks like this:

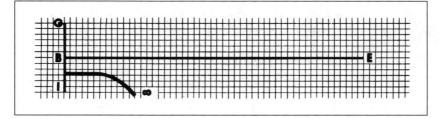

But could my graphs, when all was said and done, be useful as anything more than little visual comedies, cartoons of a sort? The University of Chicago asked me that, and I had to ask myself that, and I say again what I said at the beginning: that the graphs were at least as suggestive as pots or spearheads.

But then I had another look at a graph I had drawn of Western civilization's most enthusiastically received story, which is "Cinderella." At this moment, a thousand writers must be telling that story again in one form or another. This very book is a Cinderella story of a kind.

I confessed that I was daunted by the graph of "Cinderella," and was tempted to leave it out of my thesis, since it seemed to prove that I was full of shit. It seemed too complicated and arbitrary to be a representative artifact— lacked the simple grace of a pot spearhead. Have a look:

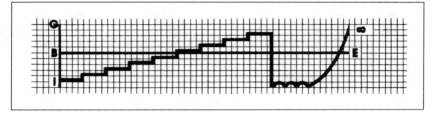

The steps, you see, are all the presents the fairy godmother gave to Cinderella, the ball gown, the slippers, the carriage, and so on. The sudden drop is the stroke of midnight at the ball. Cinderella is in rags again. All the presents have been repossessed. But then the prince finds her and marries her, and she is infinitely happy ever after. She gets all the stuff back, and *then* some. A lot of people think the story is trash, and, on graph paper, it certainly looks like trash.

But then I said to myself, Wait a minute—those steps at the beginning look like the creation myth of virtually every society on earth. And then I saw that the stroke of midnight looked exactly like the unique creation myth in the Old Testament. And then I saw that the rise to bliss at the end was identical with the expectation of redemption as expressed in primitive Christianity.

The tales were identical.

I was thrilled to discover that years ago, and I am just as thrilled today. The apathy of the University of Chicago is repulsive to me.

They can take a flying fuck at the mooooooooooooooooooon.

## Sylvia Townsend Warner

In her 1927 novel *Mr. Fortune's Maggot* Sylvia Townsend Warner writes an extended scene in which the well-intentioned Mr. Fortune, a missionary on a Pacific island, distinctly does not employ constructivist teaching methods as he attempts to convey his own mathematical pleasures to an island boy. Students might use this short excerpt as a model for constructing stories of their own learning, using the narrative elements of character, setting, dialogue, and so forth. Notice the way the thinking of Warner's characters is made evident. In the third-person-limited approach, we are allowed to see the workings of Mr. Fortune's mind directly as he imagines the success of his teaching. We see only from external evidence that Leuli is not trying to think on his own but to read Mr. Fortune's mind to figure out what he wants. Leuli's actions speak louder than Mr. Fortune's words, in the end. Although Leuli is failing in this well-intentioned but unfortunate chalk-and-talk classroom on the beach, he would do well in any of the classrooms discussed in this book. I particularly see him sailing with David Hardy under the sturdy bridges of applied mathematics.

### From *Mr. Fortune's Maggot*

Walking up and down the beach, admiring the surface which to-morrow with so much epiphany and glory was going to reveal the first axioms of Euclid, Mr. Fortune began to think of himself as possessing an universal elixir and charm. A wave of missionary ardor swept him along and he seemed to view, not Leuli only, but all the islanders rejoicing in this new dispensation. There was beach-board enough for all and to spare. The picture grew in his mind's eye, somewhat indebted to Raphael's Cartoon of the School of Athens. Here a group bent over an equation, there they pointed out to each other with admiration that the square on the hypotenuse equalled the sum of the squares on the sides containing the right angle. . . .

By the morrow he had slept off most of his fervour. Calm, methodical, with a mind prepared for the onset, he guided Leuli down to the beach and with a stick prodded a small hole in it.

"What is this?"

"A hole."

"No, Leuli, it may seem like a hole, but it is a point."

Perhaps he had prodded him a little too emphatically. Leuli's mistake was quite natural. Anyhow, there were bound to be a few misunderstandings at the start.

He took out his pocket knife and whittled the end of the stick. Then he tried again.

"What is this?"

"A smaller hole."

"Point," said Mr. Fortune suggestively.

"Yes, I mean a smaller point."

"No, not quite. It is a point, but it is not smaller. Holes may be of different sizes, but no point is larger or smaller than another point."

Leuli looked from the first point to the second. He seemed to be about to speak, but to think better of it. He removed his gaze to the sea.

# Dale Worsley

I inadvertently wrote a half-dozen poems with math themes once I started research for *Teaching for Depth*. Once mathematical ideas and models are in your mind, it's difficult not to include them in your work. It seems a natural process. Here are two of them. The first, "Shoulder Bag," came from a freewriting exercise about my Land's End shoulder bag, which that day happened to contain Morris Kline's opus *Mathematics in Western Culture*. It is loosely structured as an ode, a form in which poets are often found addressing the objects of their praise. Students might write odes to anything, including objects or ideas from their math classes.

**Shoulder Bag**

Black bunch of canvas, reality itself
strapped to a two-legged post—
inside your odiferous caverns
a whole shadow world turns on
a phantom axis.
Greek mathematicians
plot the course of the planets in there.
Neglecting their own experience,
Pythagoras and Euclid and Archimedes
walk around
the equator talking triangles,
to the north a fence of pens,
to the south ice caps of paper.

Shoulder bag, you are a crumpled burden.
when I need rest and
want to just forget who I am.
But if I were at land's end
looking out at the white caps
and I saw you
floating on the sea, a battered freighter,
I would swim to your rescue
and tow you home.
I would dry you out
and empty your hold of lost treasure:
antacid, eyeglass cleaner,
unmarked envelopes, grimy Trident gumsticks,
and a little book of Zen sayings,
including the one I never forget:
"Which is more dangerous: success or failure?"
The day I'm free of you,
black demon on my back,
I'll be free of all my mortal coils.

On April 11, 1999, the New York Knicks basketball team had trounced the New Jersey Nets by a score of 93–78. Nets point guard (a term that might conceivably have a different meaning on a court of Cartesian coordinates) Stephon Marbury was badly outplayed by the combined efforts of Knicks guards Chris Childs and Charlie Ward. After the game, Marbury was interviewed in the *New York Times* (2/17/01). The following is a poem I "found" in the Marbury interview as a result of my heightened math awareness. Students might also scan their environments not only for verbal found poems but for found mathematical visual art, as well.

**Point Guard: A Mathematical Found Poem**

They still need a point guard.
They'll never win a championship
with those two guys.
Never, ever, never.
Never continuous to eternity,
decimal, decimal, decimal.
They might not even
make the playoffs.

Note not only the mathematical imagery in this poem, how it pours poetic salt into the wound of the emphatic insult, but the inadvertent appropriation of rhythm from this line from King Lear: "Never, never, never, never, never." I am reminded also of the narrator's emphatic declamation in Dostoyevsky's *Notes from Underground*: "Are you laughing again? Go ahead, laugh, but I still won't say that my belly is full when I'm hungry; I still won't content myself with a compromise, with an infinitely recurring zero just because it is allowed to recur by some law, just because it's there. . . . "

# Alan Ziegler

Alan Ziegler uses the form of a recitation in this prose poem that intentionally conflates the highlights of what one remembers from school. Students might enjoy doing the same thing about their schooling in general, or write in a stream everything that comes to mind about mathematics. Note the right-justified formatting of the poem and the effect it produces on line endings.

**What Did I Learn In School? (A Recitation)**

Am is are was were be being been has have had do
does did shall will should would may might must can could
sine cosine cosine sine cosine cosine sine sine to be or not to
be that is the question whether 'tis nobler in the mind to suffer
the slings and arrows of outrageous force to the fulcrum force
to the fulcrum pie are square every good boy does find
almighty God we acknowledge our dependence upon pie are
square force to the good boy was were be being been nobler in
the am is are we sine cosine force whether 'tis nobler in the
mind to suffer the dependence

# Contributors

**John S. Berry** is Professor of Mathematics Education at the University of Plymouth, England, and is Director of the Centre for Teaching Mathematics. He has a B.Sc. in mathematics and a Ph.D. in applied mathematics. He has been active in mathematics education for over twenty years with research interests that include developing students' problem-solving and mathematical thinking skills, the development and understanding of key concepts, and the use of technology and symbolic algebra in teaching and learning mathematics. Away from academic life John enjoys time with his family, traveling, and playing golf.

**Virginia Cerussi** is a mathematics teacher, department chairperson, math league moderator, and freshman mentor at Loyola School, a Jesuit, independent, coed, college-preparatory school in New York City. Previously she taught at the Academy of Mount Saint Ursula, where she conducted the research for her article on journal writing in the mathematics classroom. She subsequently became assistant principal, but her strong desire to return to teaching prompted her move to Loyola in 1999. Virginia lives with her husband in the Bronx and is the mother of two grown sons. In her spare time she enjoys walking, reading, listening to music, and attempting the *New York Times* crossword puzzles.

**Peter Dubno** is currently an eighth-grade mathematics teacher at the New York City Lab School for Collaborative Studies. He is fifty-three years of age and has been teaching mathematics for thirty years in the New York City public school system. "If I'm not having a good time teaching, then I'm doing something wrong" is his workday motto. A family consisting of his lovely wife and two terrific children keep him an active man. His avocations are fly-fishing and working on the family cars. He is a member of Trout Unlimited, a conservation organization.

**Elizabeth Fox** is a writer and educator who lives and works in Brooklyn. She writes prose poems, lyrical essays, and short stories based on imaginary landscapes, strange people, and New York City. She is the author of *Limousine Kids on the Ground*, a collection of prose poems and lyrical essays. Over the past twenty years she has worked to help students at all levels from kindergarten through college enjoy writing and to help teachers enjoy teaching writing. In 1998, she helped design and lead the New York City Outward Bound© Center and Expeditionary Learning Outward Bound©'s first six-day Poetry Expedition for educators.

**Sylvia Maria Gross** grew up in Brooklyn, Brazil, and Bethesda, Maryland. She went to Walt Whitman High School in Bethesda (where she had Susan Wildstrom as a math teacher) and attended Yale University. After college, she researched the use of arts in education in Bahia, Brazil, through a Fulbright grant. She now teaches math, literature, and Bronx Studies in the middle school at Mott Haven Village School.

**David Hardy** is a high school technology education teacher in a county close to New York City. He is also the director of the not-for-profit corporation Building Bridges, Building Boats, which is dedicated to bringing people together through the activity of building and using small boats. He can often be found on the Hudson River in a homemade wooden boat floating around with various kids and adults.

**Ian Hauser** is currently a humanities and literacy consultant in elementary, middle, and high schools throughout New York City. Before moving to the United States in 1998, he was a classroom teacher, staff developer, and, for fourteen years, a school principal in Australia. He has been an adjunct professor of education at Dowling College, Hofstra University, and New York University. He has an enduring interest in curriculum development and a strong belief in the constructivist approach to teaching and learning. When not poring over the latest research, he is happily making and restoring furniture in his basement shop in Brooklyn.

**Avram J. Kline** earned his B.A. from Sarah Lawrence College and his M.A. in English Education at Teachers College, Columbia University. He has taught English and humanities at The New York City Museum School, a public school in Manhattan, since 1996 and currently participates in New York City Community School District Two's Mentor Teacher Program. Avram takes pleasure in directing student Shakespeare performances, coaching soccer, and auditing his colleagues' math classes.

**Susan H. Picker** has worked as an educator in the New York City public schools for more than twenty years, first as a teacher, then as staff developer, and now as Coordinator of Middle and High School Mathematics for Community School District Two. She holds an M.A. from Columbia Teachers College and a Ph.D. in Mathematics Education from the University of Plymouth in Devon, England. Besides enjoying trips to Wales, England, and France with her husband, she is a Brooklyn Cyclones fan who is interested in nifty gadgets, discrete mathematics, and the history of New York City.

**Kay Rothman** teaches humanities to sixth graders at the New York City Lab School for Collaborative Studies in Manhattan. Previously she made her way as a painter of portraits during and after more than ten years of waiting tables. She earned her B.A. in English from Indiana University, her M.S. in psychology from San Francisco State University, and her M.S. in Museum Education from Bank Street College of Education. Kay enjoys teaching, her family, her friends, reading, theater, movies, and researching curriculum in varying degrees at various times.

**Amy Samson** teaches English at North Eugene High School in Eugene, Oregon. She earned her B.A. in English literature and studio art at Skidmore College and her M.A. in English education at Teachers College, Columbia University. She has taught in the People's Republic of China, in the New York City public schools, and at Philips Academy at Andover and currently teaches and resides in Eugene. In her classroom, she guides her students to explore connections among writing, literature, visual art, science, and mathematics. Outside the classroom, she enjoys backpacking, photography, reading, and culinary adventures with her husband.

**Matthew Szenher** is an adjunct lecturer at Hunter College in the department of computer science. He has an Sc.B. in computer science from Brown University. Prior to his stint at Hunter, Szenher taught computer science and mathematics for four years at The Dwight School in Manhattan. When not in the classroom, Szenher pursues his interests in long-distance running, Civil War biographies, and experimental cookery.

**Tammy Vu** is currently assistant principal at Boston Evening Academy, where she focuses, among other things, on developing the math program. She has taught secondary math in a variety of schools, from a polytechnic in Singapore to a large Catholic school in Queens to a small alternative public school in Manhattan. She attended Princeton University and recently received her master's degree in education from Harvard University.

**Matt Wayne** teaches middle school language arts at a small middle school in New York Community School District Two. He is also a MetLife fellow in the Teachers Network Policy Institute, an organization committed to bringing the teacher's voice to policy making. Matt continues in his role as language arts teacher at home, reading to his one-year-old daughter, Ella, every night.

**Dale Worsley** teaches writing at Columbia University and works as a curriculum and literacy consultant for school districts and arts organizations. He is the author of the educational resource books *The Art of Science Writing* (coauthor) and *Portfolio of a School*. Awards for his writing include National Endowment for the Arts and National Endowment for the Humanities grants and EDPRESS awards. He is in the process of trying to interest his six-year-old daughter Elly in bird-watching and basketball.

# Index